THE ACCIDENTAL NETWORK

THE ACCIDENTAL NETWORK

HOW A SMALL COMPANY SPARKED A GLOBAL BROADBAND TRANSFORMATION

Rouzbeh Yassini-Fard

WITH STEWART SCHLEY

FOREWORD BY JOHN CHAMBERS

WEST VIRGINIA UNIVERSITY PRESS
MORGANTOWN

Copyright © 2025 Rouzbeh Yassini-Fard
All rights reserved
First edition published 2025 by West Virginia University Press

Printed in the United States of America

ISBN 978-1-959000-60-0 (hardcover) / ISBN 978-1-959000-61-7 (ebook)

Library of Congress Control Number: 2025009550

Cover design by Michel Vrana
Book design by Than Saffel

NO AI TRAINING: Without in any way limiting the author's exclusive rights under copyright, any use of this publication to train generative artificial intelligence (AI) technologies to generate text is expressly prohibited. The author reserves all rights to license uses of this work for generative AI training and development of machine learning language models.

For EU safety/GPSR concerns, please direct inquiries to WVUPress@mail.wvu.edu or our physical mailing address at West Virginia University Press / PO Box 6295 / West Virginia University / Morgantown, WV, 26508, USA.

*For my mom, dad and uncle. And my sister Pamela.
Thank you for believing.*
—Rouzbeh

*Out beyond ideas of wrongdoing and rightdoing,
There is a field. I'll meet you there.*
—Rumi

CONTENTS

Foreword by John Chambers xi
Prologue xiii

1 American Dream 1
2 Wired Nation 12
3 Vision Quest 23
4 It's a Deal 41
5 The Missing Link 52
6 When Doves Fly 67
7 Death Star 80
8 Puppy Love 93
9 Screen Wars 103
10 Rose Gardening 112
11 Wicked Fast 126
12 Father Figure 151
   Epilogue: A New Alphabet 165

Acknowledgments 171
In Gratitude 173
Notes 175
Index 183

FOREWORD

Some of us may recall the early days of connected computing, where the squeal of a phone line modem and a sometimes-agonizing wait for text, images, and multimedia content hinted at exciting new innovations. It was during this era of the mid-1990s when I predicted that "voice will be free" and that all forms of communications would move to IP. The concept of free voice service was heavily contentious at the time, given the sluggishness of Internet connectivity and the fact that voice was the main source of revenue for telephone companies. Fast-forward, and we see both ideas—free voice service and the widespread migration to IP—have prevailed. Not only that, the idea of manually "going online" or "connecting to the Internet" has all but vanished, replaced by an environment of always-on connectivity.

The critical enabler of this transformation is "broadband," a catch-all term describing high-speed, reliable, and secure transmissions that now underpin the fabric of modern life. It is hard to convey the scope, breadth, and impact of today's broadband networks, but, in short, broadband has completely changed the way we work, live, learn, and play. What's more, broadband has empowered innovation. It has invited startups and established companies to share and implement game-changing ideas and intelligence with a scope and scale we never thought possible before.

A convergence of many inventions and technologies has made all of this—our modern idea of the Internet and human connection—possible. For me, two in particular stand out. One was the establishment of an Internet-routing architecture that harnessed the power of IP. The other, which is the core focus of this book, was the "cable modem." This device transformed a networking infrastructure built for television viewing into a scaled mechanism for high-speed data transmissions, empowering everyday citizens. Both were and still are astonishing, breakthrough technologies that created extraordinary value.

Behind every technology story is a "people" story that encompasses familiar themes: the power of the human drive and the ability to anticipate market transitions. *The Accidental Network* captures both the technological and human stories of how we have gotten to where we are—and how we will keep pushing the envelope to evolve the Internet beyond what we can even imagine.

John Chambers
Founder and CEO, JC2 Ventures
Former Executive Chairman and CEO, Cisco

PROLOGUE

The Accidental Network recounts the improbable story behind the invention of the cable modem, a breakthrough technology that has supercharged today's broadband revolution.

Broadband is the Internet at warp speed. It's the foundation for the anything, anywhere, anytime economy, a pipeline that connects people, organizations, machines, and communities, and in doing so, touches nearly every aspect of our daily experience—how we communicate, work, learn, are entertained, receive health care, apply for jobs, monitor our physical environment, buy goods, access podcasts, just about . . . anything. Broadband has changed the fabric of everyday life, connecting billions of people, powering the ambitions of Silicon Valley innovators, and creating connectivity that enables individuals to thrive as producers and creators, not merely consumers.

Today a large share of broadband traffic flows over the infrastructure once known as "cable television," one of the world's most prolific broadband delivery mechanisms.

It wasn't supposed to happen this way. The cable industry originally was created for one purpose: to distribute television channels. First in hilly hinterlands and later across suburbs and metroplexes, cable companies brought television signals to millions of people. Early cable pioneers started out in the late 1940s by extending the reach of local, over-the-air TV stations. Later, they opened the floodgates for entirely new channels—familiar television networks like ESPN, The Discovery Channel, and CNN.

By the early 1980s, the cable industry had emerged as a bona fide entertainment juggernaut, attracting investments from giants of the media and communications establishment. Companies like the publisher Time Inc. and the manufacturing giant Westinghouse Electric Corp. invested heavily in cable television operations. Entertainment conglomerates, including Walt Disney Co. and Warner Communications, launched flashy new cable TV channels.

But then, pressure. A 1992 law imposed severe economic strictures, forcing providers to roll back prices and watch revenues tumble while bankers and investors recoiled. Competition from a satellite-powered pay television system began to tear at cable's flanks, peeling off millions of customers. And behind it all, a new phenomenon was taking shape: Online services like CompuServe, America Online, and Prodigy were introducing new ways to communicate over connected computers, siphoning attention from the television screen entirely.

With the world changing, the cable television industry needed a second act. And I happened to have one in mind.

In the residential cable pipelines that coursed across the country, I saw something more than a delivery system for entertainment. I saw the makings of a converged, high-speed data delivery mechanism that could forklift the power of the high-speed local area network from the office to the home, yielding a holy trinity of unlimited users, long distances, and high speed.

In 1987, I became convinced we could leverage the existing infrastructure of residential cable television to make the city-wide broadband idea real. Cable didn't have to be just about television. It could also be the underpinning for the long-awaited information superhighway.

I had a vision, and I was determined to make it happen. In 1990, after two years of trial deployments and technical research, I assumed control of a struggling Massachusetts technology firm whose first-generation product attempted to connect data terminals over long distances using private cable systems. It was heavy, boxy, unwieldy. But it was a start.

Timing turned out to be on our side. The cable industry, eager for renewal, was looking to stretch its wings. Cable companies began adopting the language of digital communications to free up capacity and launch new services. They transformed their infrastructure to allow for two-way communications. And one by one, they began doing business with a small company from a Massachusetts mill town.

This is the story of that pioneering company. In telling it, I'm very much aware that broadband did not spring to life purely because of a vision that came to me in 1987. Broadband happened because of the contributions of many. And in some ways, it happened by virtue of some lucky accidents. But as everybody knows, accidents can change the world.

THE ACCIDENTAL NETWORK

CHAPTER 1

AMERICAN DREAM

You hustle. You persevere. You lose sleep.

Icy strands from a weekend storm clung to trees outside our fourth-floor office in Andover, a Massachusetts mill town edged up against the Shawsheen River. The parking lot outside was covered with hardened snow. Car tires pressed over a snow-crusted surface as early arrivals steered their way into empty parking spaces.

It was Tuesday, December 8, 1992. I had stayed at the office overnight, typing out a year-end accounting summary for our tax advisor. It gave me something to do during the restless early morning hours as I awaited this morning's delivery.

By now, skipping sleep was the norm. Staying awake was a way to maximize my contribution, multiplying my productivity two or three times over. I was president, chief technology officer, keeper of the flame, occasionally office janitor. My sleepless record to date was 73 hours straight. Not entirely healthy, but . . . effective.

The good news was that we were getting close. I could feel it. Something had shifted in the last week or so. People around me were scurrying about a step faster than usual, including my tousle-haired quality manager, Mike Sperry, who was among the first to arrive that morning. I hustled over to greet Mike in our engineering documentation room before he had a chance to unzip his winter jacket.

Mike and I were waiting anxiously for an overnight package containing a set of delicately fabricated integrated circuits—thin wafers sprouting tiny copper legs. Once we had them in hand, it would take Mike about an hour to drive to Westborough, Massachusetts, where he would deliver our trio of chipsets to our manufacturing partner. There, they would be delicately nestled into electronic circuit boards, with 160 tiny pins—40 per side—aligned with their receptacles. We were scheduled to get the boards back within 72 hours. We would run them through a rigorous progression of demanding technical tests over the coming weeks.

And then we'd know.

It had been 30 months and 20 days since I'd pledged my house in nearby Boxford as collateral to the lead investor of a company I was working to acquire: a struggling tech outfit with a faint glimmer of upside in the form of a heavy, expensive communications modem.

Back then, my house was all I had to offer. My bank account wasn't impressive, and hardly anyone, even some of my own colleagues, fully believed in the dream. Venture capital investors and banks wouldn't touch us. My company was bootstrapping it the whole way, eking out payroll, sweating every check, bargaining with vendors to loan us pricey test equipment for a month or two at no cost. We were deep into the great American entrepreneurial dream, familiar to anyone who has ever started a business. You hustle. You persevere. You lose sleep.

What hung in the balance was an audacious idea: a 24/7 connected society where everyday citizens could interact, learn, do business, and create, geography notwithstanding. Where data streams would zip across entire cities at speeds once confined to corporate office buildings. Where the large-scale deployment of residential broadband networks promised to deliver incalculable value—trillions of dollars in market capitalization, a flood of entrepreneurial creativity, a new alphabet for human invention.

Since 1987 I'd been obsessed with the idea of a networked city and how to make it real. But getting others to buy in was still difficult. Although a handful of cable industry technologists had been talking about the possibility of data transmission and computer industry alliances, much of the energy and investment in the cable business in the early 1990s was still focused on the familiar practice of delivering television channels. The industry research and development organization CableLabs was only beginning to probe at the edges of a new model for service delivery. The group had scheduled its first management briefing on "multimedia services" for industry executives in April 1993.

We were ahead of the game, having worked to make the case for several years. A 1991 marketing brochure I commissioned used a comic-book style to try to explain the promise of broadband data transmission. One frame featured an office worker chatting on a desktop telephone while typing away at a keyboard. "Good morning, boss," chirped the character in an illustrated thought balloon. "I'm reporting to work from my home office 35 miles away.

I just completed the report in record time and am sending you the finished copy as we speak."

That scenario is now commonplace; today the comic reads as quaint nostalgia. But in 1991, it was still unfamiliar terrain, part of an "information age" promised in popular 1980s books like Alvin Toffler's *The Third Wave* but absent from everyday experience. The reality was that the post-industrial information age was barely tottering to life. Telecommuting with any degree of scale was the stuff of fantasy. The connected classroom remained a distant dream. The Internet was mostly text messages and bulletin boards used by government and military researchers and academics.

The prototype chipsets that were on their way to our office were supposed to change that. They encapsulated nearly five years of intensive development work we had done the New England way: taking a hardened block of granite and working the edges, methodically yielding the contours of a technology that might just light up the world. Along the way, my team confronted hundreds of seemingly oppressive technical challenges. We chipped away.

It often made for rough days. To keep our spirits up, we celebrated every advancement. We marked our progress with high fives, plenty of peer recognition, dart games in the kitchen, and homemade Persian dinners cooked by my mom. Now, here was one more small victory standing right outside our door: The FedEx guy had arrived.

* * *

1968. A weekday afternoon. Ten-year-old me. Curled across the bright green bicycle frame is the insignia of the British bike maker Hercules, named after the hero of Greek mythology. But the rough-and-tumble brand seems out of place. For one thing, I'm Persian, not Greek. For another, it's a girl's bike, and I'm a boy. Like every kid growing up around Tehran, I had aspirations of style, of being seen as a cool young man, determined to grow a mustache one day. But there is no way to present the desired image with an expandable handbag—a purse, for goodness' sake—strapped to the handlebar.

Still, it's the only bicycle my family owns, and when I hop onto the thing, pressing the thick black pedals with as much force as I can, it's with a sense of resigned determination. My mom needs groceries, and with no refrigerator in our tiny kitchen, we can store only so much food on any day. Jaunts to the market are a fact of life. From my early childhood, I internalized a sense of responsibility and an indestructible bond with my mom.

So sometimes in the darkness of early morning, sometimes after school, I make the half-mile, roundtrip ride over a hard-packed dirt road to the shops on Street No. 2 in the neighboring community of Nirouhavi, where the owners of a grocery store and a small baked-goods shop are faithful enough to extend us credit on days when the money is short.

And money is frequently short. Often, I return home with only a stained paper bag containing chicken legs, the centerpiece ingredient of a thin soup that will feed me and my family: my parents, my older sister, my older brother, and the baby of us all, Pamela.

We eat modestly. We get our drinking water by refilling buckets from a water carriage. But we don't think of ourselves as poor. Life in our house in the small town east of Tehran is simply what we know. My siblings and I sleep close to one another, laying out a homemade mattress over a thick rug and curling up in the hallway leading to the house's lone bedroom, where my parents sleep. In the morning, we roll up our gear, tuck it into a corner of the bedroom, and begin our day. Evenings, we gather in the same room, place a colorful plastic covering over our rug, and have dinner.

Now, fast-forward to a September afternoon in 1996. I'm sitting in a leather chair in a downtown Boston conference room, headquarters for the law firm Ropes & Gray. Wearing my customary suit-and-necktie uniform, I sign my name across what seems to be a bottomless stack of papers embedded inside a stack of five or six thick binders. I'm completing the sale of a business that a handful of colleagues and I had resurrected nearly from ground zero. The sale price is $59 million.

From the remnants of a predecessor named Applitek Corp., we had achieved the improbable. Surviving on a bootstrapped revenue stream, delaying my paycheck and my younger sister's paycheck to make payroll, persisting when some of the top minds of the computer and telecom industries told us we were doomed to fail, my team was grinding away to pursue a vision for changing the world. Via the invisible hand of a wire transfer, our bank accounts had swollen to an amount my 10-year-old self, pedaling hurriedly to fetch groceries on weekday afternoons, could never imagine.

One of my colleagues from the cable industry used to refer to me as "the crazy Persian." David Fellows, an exceptionally bright technologist, had a point: My approach tended to run outside customary lines. And it's true we worked like crazy. Late evenings at our old office in the former mill town of Andover were the rule. My mom prepared lavish Persian dinners for our team: plates overflowing with steaming Persian stews—her eggplant khoresh a team favorite—and delicacies like grilled kabab koobideh, rice with hints of saffron and pistachio adding a savory kiss. For mom, the meals were an expression of love. For me, they were a way to keep my troops fed and happy—and to coax people into sticking around for a final burst of work before heading out after 9 p.m. or so. They also preserved a connection to my native country, where centuries before, the kitchens of Persian empires were famous for their elaborate feasts. I kicked off our communal meals with the expression "nush-e-joon." It translates to "food of life." The phrase seemed appropriate.

I shared the proceeds from the company's sale with my extended team. They had championed the deal, mindful of mortgages and tuition bills to pay. But more important than the money was the achievement. With our smallish start-up now in the hands of a bigger, more resourceful parent, our dream of connecting the world, empowering everyday citizens, and making a positive impact on the physical environment seemed likely to happen.

I shook hands with my attorney, David Fine, along with a roomful of lawyers and advisors. I returned down the elevator to the parking lot. It took an hour to navigate the drive back to our new office, a nondescript collection of smallish cubicles inside a single-story suite on Bullfinch Drive, not far from the Merrimack River. Before I'd exited the room, Fine had pulled me aside as we concluded the transaction: What was I going to do now? I think he half-expected me to celebrate with a trip to a tropical island. Not so much; I told him I had a commitment to ship out orders and to finish up details on a new technical specification.

I was back in my office by early evening. I settled in at my desk, the last tickle of an eastern sunset my companion. I brewed a cup of Earl Grey tea and booted up my desktop computer.

Our ambition was to transform the nation's cable television systems, conceived in the late 1940s to let rural Americans watch TV shows like Milton Berle's *Texaco Star Theater*, into a medium that would change the way people lived, communicated, worked, and played. A medium that—I was certain of this—could make a meaningful dent in the world's seemingly unquenchable thirst for oil. A medium that, years later, would help the global economy survive during the ravages of a worldwide health pandemic.

Our mainstay product was the cable modem. We had spent close to eight years developing our technology, and by 1996, we were gaining momentum, crisscrossing the nation and a growing list of international locations with new deployments.

A few years before, venture capital investors I'd met with told me the idea of a vast, high-speed residential data network connecting millions of users with cable TV lines was a nonstarter. They were convinced that the "cable guys" would never be serious about high-speed data delivery and that I was wasting my time.

The office was quiet now. Thin wisps of steam rose from my teacup. I scanned the computer screen: glowing orange type against a dark background. We may have sold the company, but there were still orders to fill, work to do. It had been nearly 30 years since I pedaled to fetch the day's groceries, 19 years since I arrived in the United States, nearly 9 years since I'd conjured a vision of a connected society. It was the autumn of 1996. The revolution was on.

* * *

A clerk at the Pan Am building at New York's JFK Airport pointed a finger to my left, and I took off sprinting. It was close to 7 p.m. on July 31, 1977. I was running late for a connecting flight on Allegheny Airlines that would take me to Pittsburgh, where my older brother and his American fiancée—having preceded me into the United States three years before—would be waiting to pick me up.

The siren song of opportunity brought me to the United States when I was 18 years old—the shared dream of so many immigrants. That, and the specter of a deteriorating Iran: Discontent over the Pahlavi monarchy being engineered by outside powers was mounting. I had arrived in the United States only two years before the destructive regime of radical Islamic rulers seized power.

I had learned passable English while attending the Pahlavi University in Shiraz, Iran, but I struggled to keep up when people talked fast—which is exactly what was happening now. I thrust my ticket toward the attendant and, with a pleading expression, managed to ask for help. Since boarding my Pan Am flight in Tehran, I had been traveling for 17 hours, including a long delay in London. I was whipped. But I was determined not to blow this; everything depended on me ultimately arriving in West Virginia, where I had pre-enrolled at West Liberty State College near the city of Wheeling.

Somehow, I made it to the Allegheny flight, and to Pittsburgh, where my brother and his fiancée met me at the gate with a welcoming smile and a full tank of gas. By the time we reached the campus, I had arrived for good in the United States of America. Except for business trips abroad, I would never again live anywhere but in the United States, my adopted and beloved country. I haven't returned to Iran, presuming that my name is on somebody's list of unwelcome visitors: a man who had the gall to develop a technology that empowered billions of ordinary citizens.

Still, I mourned the disintegration of a country, some 3,000 years old, that had given the world so much: art, poetry, science, human rights, culture, music, heritage—and, of course, food. Much of which was now at risk of being swept away.

I was fortunate. My extended family had been able to pool enough money to support my $750-per-semester tuition. A year later, my mother and younger sister Pamela joined me, providing a welcome sense of home and family to what was still an unfamiliar place. My father would arrive in the United States a few years later.

America was breathtaking. A place where nothing stood still, where hustle and verve ruled. A land of inexpressible physical beauty. A country where hard work mattered, where a janitor could become a CEO. I was grateful for it all, especially for an American tradition my fellow citizens called Thanksgiving. Something about this blend of food, friends, family, and ritual captivated me. It was the holiday everyone seemed to look forward to—football, parades, family gatherings, and the meal itself, the ultimate celebratory and culinary experience. Beginning with our inaugural Thanksgiving in 1977, my late mom and I ended up hosting more than 40 Thanksgiving gatherings, bringing students, immigrants, colleagues, family, and friends together with the goal of helping those new to American culture embrace a tradition that cultivates respect for our country.

Since arriving in the United States, I've worked with hundreds of students to help them understand and embrace the American dream and the traditions undergirding a remarkable nation, including, of course, sports. Shortly after my arrival, I became a card-carrying fan of the Big 3 Pittsburgh teams: the Penguins of the National Hockey League, the mighty Steelers of the NFL, and the Pirates of Major League Baseball. Watching games on TV from a campus lounge called the Mountainlair, where I'd gather with a few engineering students, fill a cup with hot tea, and settle down to study while sneaking glances at the game, remains an enduring memory.

Not every American custom came easy. Although my English steadily improved, I missed certain signals. I suffered razzing from a roommate during my freshman year at a West Liberty State College fraternity for neglecting to hew to a time-honored code: a sock wrapped around the exterior doorknob of a bedroom meant—this was not to be missed—that there was a guest in the room, and nobody was to breach the threshold. That, and about a dozen other rituals of early college: pennies lodged underneath doors to make them impossible to open; cold water tossed over the shower door to the horror of my exposed skin and the delight of my roommate. I wasn't a fraternity member, but my late arrival that fall left few dormitory rooms available. I shared a room with the frat-house leader. Welcome to college.

Other experiences weren't so innocent. The 1979 seizure of the US embassy in Tehran lorded over everything. America held its collective breath while radicals held more than 50 Americans hostage. Every evening, the ABC News correspondent Ted Koppel stoically recited the names of the Americans held captive—a sober ritual that darkened America's mood during the final days of the Carter presidency.

The international trauma distressed me. I worried for everyone back home and felt guilty for my own experience. Here I was, living in a newfound personal heaven, having traded a hard life for a comfortable community of fellow students and friends in a place where opportunity and optimism abounded. I watched from afar as the unrest worsened. In the meantime, my immediate remedy was to avert eye contact: in elevators, in hallways, in the rapid transit systems that shuttled us to and from the campus at Morgantown's West Virginia University, the college I had transferred to in my sophomore year.

My plans since my teen years had been to follow my uncle's example: become a physician. The medical field seemed infused with potential, with no end to discovery, exploration, and possibility. Becoming a doctor would be a way to contribute to humanity at large. But at WVU, something unexpected happened: I fell in love with electronics. Four brilliant professors—James Corum, Constantine Balanis, Roy Nutter, and William Cooley—exposed me to the world of computer science, electronics, electromagnetic theory, and antenna technology. I became transfixed.

It was the era of the punch card: thick paper on which computing instructions were indicated by the presence or absence of painstakingly produced holes. Alternatively, if we were lucky and a machine was available, we could type computer code directly into a terminal. To maximize my time at the computing lab within the engineering school, I devised a scheme: I'd sign up for one of the late-night slots in the hope that one or both of two outcomes would occur. The first would be that the student scheduled to precede me would give up or be lured away by the promise of a date or a pitcher of beer with friends. The second would be that the student slated to follow me would abandon the slot, deciding not to wake up and trudge across campus at 3 a.m. The result: on many nights, I'd have the lab and a PDP-11—a 16-bit minicomputer—pretty much to myself after midnight until early in the morning. After that, I'd pack up my work, nab one or two hours of sleep at my apartment, and begin another day of tutoring fellow freshman students and attending my own classes. It was not the healthiest of routines. But it maximized my time learning about intoxicating new fields: computing, electrical engineering, and the emerging communications satellite sector. As for the tutoring, I needed the cash.

In retrospect, I was probably destined for the world of electronics from the start. When I was 14, I purchased an AM/FM radio kit, a prepackaged collection of solder, printed circuit boards, and miniature components. Working in the backyard so as not to poison our house with solder fumes, I became captivated by the thing, staring and studying for hours at the circuits and marveling at the roles played by even the tiniest of elements.

My fascination deepened when, in the spring of 1981, I began working on a senior class project with four other students. Under the guidance of Professor Corum, we were charged with developing a satellite downconverter, a device that would collect radio waves from a nearby satellite dish and convert them to a frequency that could be understood by a connected TV set receiver, just like an over-the-air television signal. I was in awe. Here we were, a quartet of students living and working in Morgantown, West Virginia, devising ways to communicate with a geostationary satellite hovering in space 22,300 miles away. I began to understand the marvels of telecommunications—how we could bridge entire worlds, connect people, and reduce disparities between big cities and rural communities.

With my graduation from WVU nearing, I started a job hunt. From my tiny apartment in Morgantown, in the fall of 1980, I'd sent hundreds of resumes out the old-fashioned way: typing, photocopying, hand-addressing, and hoping. That batch of 300 or so missives produced a total of 10 interview requests and, subsequently, two job possibilities. I accepted an offer to work as an entry-level engineer with General Electric at the company's sprawling consumer electronics division in Portsmouth, Virginia, even though I knew little to nothing about the innards of TV sets or, for that matter, what they displayed. Other than my Pittsburgh sports teams, I rarely even watched television.

My proud assumption had been that my degree from WVU, my place on the dean's list, and the 3.7–grade point average I'd racked up—in the top 10 percent of my graduating class—had been noticed by some attentive company leader. Not so. Nine months later, a human relations manager confessed that GE had fallen behind on a pledge to hire minority candidates. As an Iranian, apparently, I was shuffled toward the top of the candidate pile. Not exactly an elevating moment, but at least I was working.

Besides supporting my immediate family, I was helping to raise a younger cousin—now a board-certified physician in San Diego—along with hosting regular visits from my father. GE paid me an annual salary of $13,500, a sum demanding that the five of us find creative ways to live. For the first three months in our rented townhouse on Radcliffe Lane in Chesapeake, we reprised a custom familiar from our life Iran: sleeping on mattresses without bedframes, as I saved money to buy proper furniture. On Saturdays, I mowed neighbors' lawns for extra cash.

Still, we were happy, and that was what mattered. None of us had illusions of a lavish life in America. Driving south across US Interstate 68, we invented a running family gag about an Atlas Van Lines truck we kept eyeing from behind the windshield. As we passed by the vehicle, five of us squished into my AMC Concord, we joked that it carried a bountiful supply of fine furniture and cherished belongings we would soon unpack and place in our new home. In reality, we had packed up only a few scattered items, including a classic college-era shelving system of cinder blocks and wooden planks.

The good news was that GE offered me something more important than a (barely) livable wage. I was learning from legends, individuals who had devoted their entire adult lives to their careers. Sitting at one of four desks arranged corner-to-corner along a large hallway, I was schooled by colleagues in an important lesson of consumer electronics: Whatever you do has to work on an enormous scale—reliably, efficiently, and at the lowest possible cost.

Our marketplace was all about hardware: TV sets that lorded over America's living rooms, beaming out the latest episode of the primetime drama *Dallas* or the sitcom *The Jeffersons*. I led an effort to validate a new type of TV set power supply, one that would displace heavy, expensive transformers with a cheaper, lighter alternative. I learned the mathematics of scale: how adding even a few cents of cost to a product, multiplied over millions of units, could bust apart a budget or ruin a business plan. Just as important, I met a man named Bill Corley, a GE lead engineer who was the company's resident genius of the radio frequency (RF) environment, a catchall term for systems that use radio waves to transmit information. If I ever needed a master of the RF realm (and I would), Bill would be the perfect candidate.

A year or so into the job, I was becoming a known quantity, partly because people learned they could count on me. I'd arrive at the building in Portsmouth at 7 a.m. and frequently would turn off the lights at night when I left. We weren't allowed to work on weekends at the office, so I'd often cart home a TV set on Friday afternoon and tinker with it throughout the weekend.

It didn't feel like work. I was having fun. The division where I worked at GE was a giant university, populated with brilliant engineers who were open about their work. The brainpower in that building was astonishing.

Until it wasn't.

Under the iron-fisted rule of the legendary CEO Jack Welch, GE was bent on owning the no. 1 or no. 2 market position in any business it entered, a mandate Welch had famously issued to his division managers in 1983. But by the mid-1980s, GE had fallen behind a rising cadre of Asian TV set manufacturers that competed doggedly and successfully on price and quality and were early to adopt the integrated circuit as a successor to old-school, transistor-based circuit boards.

That was enough for Welch and the newly minted, Harvard Business School-bred management at GE. Overnight, our talent was drained away. People who had worked for GE for their entire careers were jettisoned. It was a brush-up against the hard world of corporate management that I would remember for the rest of my life; GE terminated hundreds of engineers, manufacturing experts, product specialists, and managers. I watched, astonished, as individuals I considered to be intellectual giants were cast aside. I had been fortunate to learn from experts with 30 and 40 years of experience, and it seemed to me there would be no way to reclaim the knowledge base.

Certainly not by hiring the "penguins." That was the name my surviving colleagues and I applied to the bespectacled, suit-and-tie collective of MBA-endowed successors GE was counting on to save us. They were perfectly pleasant young men—congenial, self-assured, educated, and intelligent. But to us, they all looked, talked, and acted the same. Several of them drove identical Chevrolet Corvette sportscars gifted by GE as a signing bonus; six or seven were lined up like dominos in the parking lot every morning. But from what we could discern, not one of the penguins knew anything about television sets, or engineering for scale, or how to design better consumer electronics products. Instead, their idea was to stop making TV sets altogether and instead resort to private-labeling TVs made by Korean manufacturers. After that strategy failed, GE combined its consumer electronics group with that of a former rival, RCA, eventually selling the combined business to France's Thomson S.A.

I took careful stock of what had happened—the dismissal of talent, the wasting of resources. Earlier, GE had enrolled me in its Financial Management Program, an MBA-style curriculum where I learned about corporate finance, accounting, and management principles. But the real education came from watching a company do many things the wrong way. I tucked my newfound knowledge away, certain it would be useful someday.

CHAPTER 2

WIRED NATION

Even in the early days, there was a sense that cable systems might be able to do more than just transmit TV channels.

Geraldine Laybourne was three years old when a TV installer plugged in a new television set at the Laybourne household in Plainfield, New Jersey, a bedroom suburb of New York City. The youngest of three sisters, she was convinced the new medium worked both ways: You could watch characters on the screen, and they could watch you. Fittingly, she dressed in costumes before the TV set, in the faith that the TV cowboy Hopalong Cassidy would notice.

Thirty years later, her fascination with television lingered on. In the winter of 1979, after jobs in teaching, media studies, and research, Laybourne and her husband began producing television shows for a kid-friendly TV channel being run out of a smallish studio in Columbus, Ohio. The channel's originator was a division of the big media conglomerate Warner Communications, which doubled as the Columbus cable television provider. Under the moniker of QUBE, Warner was conjuring ideas for futuristic interactive television services, alongside more traditional television programming. Laybourne accepted a job as a program manager, and the team came up with a sprightly name: Nickelodeon.

The novel kids' channel was one of dozens that would work its way onto the media scene in the early 1980s. Alongside Laybourne's Nickelodeon rose a sister channel from Warner called MTV: Music Television. Dozens of other newcomers like The Weather Channel, Bravo, Country Music Television, The Discovery Channel, and USA Network also sprang to life.

All of these channels were spawned by a recalibration of how television worked. The legacy television world depended on the transmission of signals over airborne radio waves. TV stations affiliated with national networks like CBS and NBC were beamed to rooftops and "rabbit ear" antennas. These invisible waves lived within an assigned spectral range—a swath of electromagnetic energy. Because of the way the US Federal Communications Commission (FCC) parceled out this spectrum to minimize interference, there were limits on the number of television stations permitted to transmit in a given geography.

Cable television was different due to the unique conduction properties of the "coaxial" cables that gave the medium its name. The "coaxial" label refers to the pairing of an interior copper wire with a surrounding metal casing. The gap between the centered wire and the outer tube produces tremendous spectral bounty—a shielded, protected space that can operate at high-frequency ranges, expanding the boundaries of television. In these sturdy copper wires, the industry possessed a powerful force for distributing a torrent of electronic information over long distances.

Despite the distinctions, there was an important kinship between cable television and old-school, over-the-air broadcasting. It related to the FCC's scheme for meting out slices of spectrum over which TV signals could travel. In prescribing the rules for television broadcasting in a 1952 Report and Order, the FCC codified a frequency swath of 6 megahertz (MHz) for each television channel. And there it was: the prescription for a data revolution. Decades before anyone had ever heard of the "information superhighway," the FCC had enshrined the lanes of travel for the broadband data network of the future, even if nobody knew it at the time.

Instead, during the 1980s, as Laybourne's Nickelodeon grew and flourished, it was still all about television. And momentum. The FCC's annual report about the cable television industry pegged total revenue in 1982 at $3.5 billion—up a lofty 60 percent from what it had been the previous year. To the astonishment of the Big 3 broadcast TV networks—ABC, CBS, and NBC—the cable industry was redrafting the contours of the medium, with millions of new subscribers being added every year. Fresh TV channels like CNN and HBO were appealing to TV-hungry viewers who couldn't wait for cable lines to be lit up in their neighborhoods. Industry marketers like United Cable Television's Charles Townsend embraced the term "truck chasers" to describe eager early adopters "who used to knock on the doors of our service vehicles to get them service ahead of everybody else."

And chase they did. Early in 1980, more than 15 million US homes had cable TV connections—roughly one in every five. Five years later, the number had swelled beyond 35 million—more than one in every three. By 1988, not long after I had an epiphany about what cable television might do for the world of data delivery, the industry had reached a symbolic tipping point, when more than half of the homes in the United States were connected to cable TV.

Just as important as how many people were subscribing was where they lived. By the mid-1980s, cable television was becoming available just about everywhere, its presence across rural towns, suburbs, and big cities alike creating a far-reaching, nearly ubiquitous means to connect a nation.

It was a bold leap from where the industry had been. CATV (for "community antenna television") had begun life in the late 1940s as a way to extend over-the-air TV signals to rural areas where rooftop antennas couldn't rein in signals. Debate persists over where the first community antenna system was fired up, with votes split between Mahanoy City, Pennsylvania (via an appliance store owner named John Walson), and Astoria, Oregon (courtesy of a radio-station engineer named Leroy "Ed" Parsons). Regardless of provenance, however, by the mid-1980s cable television was expanding its footprint to include more populous suburban and metropolitan communities. This was critical. Had CATV remained a medium for the hinterlands, the "accidental network" might have never materialized. Instead, the realization that cable television could be a moneymaker just about anywhere drove the industry to build out systems in suburban neighborhoods, in cities dense with apartment buildings, and just about all points in between.

The pent-up demand translated into serious money for those who were quick to the game. In 1982, in Montgomery, Alabama, a 20-year-old college student named Jim Mott raked in more than $30,000 in one year (roughly $97,000 in 2025 dollars) selling cable TV door-to-door for Storer Communications, a former radio broadcasting company that was now all-in on cable. At a time when his peers were working minimum-wage summer jobs at $4.25 per hour, the former door-to-door Bible salesman was pocketing serious money: "A truckload," recalled Mott. "More than a guy who was twenty years old ought to have." Like many young people who fall into a burgeoning industry, Mott had found his calling. He would go on to become a general manager of multiple cable systems in the eastern United States.

Few individuals rode the cable wave more zealously than a swashbuckling Georgia entrepreneur named Ted Turner, whose June 1980 launch of Cable News Network (now CNN) underscored television's wide-open possibilities. The mustachioed Turner was one of the first to seize upon the power of communications satellites for hurtling signals down to the receiving stations of local cable companies and then traveling across amplified cables to millions of US homes.

Thanks in part to Turner's satellite epiphany, new channels piled on—names like Arts & Entertainment Network, CNN2 (later, Headline News), Lifetime, VH1, and dozens more. By 1989, the National Cable Television

Association, an industry trade group, counted 79 cable channels—nearly triple the number in the market at the start of the decade. "Television was being cannibalized—and democratized," observed the writer Ken Auletta in *Three Blind Mice*, a 1991 chronicle of the pressures facing the dominant TV networks ABC, CBS, and NBC.

Throughout the 1980s, times were good in cable land for a close-knit collective of leaders, including the patriarchal captain of industry, John Malone, the CEO of a rising cable TV systems operator named Tele-Communications Inc. (TCI). A graduate of Yale University (electrical engineering and economics) and Johns Hopkins University (industrial management), Malone had worked for AT&T's research and development arm, Bell Labs, before honing his skills as an executive for the cable technology supplier General Instrument Corp. Malone brought a sharp eye for deal-making as he went about buying up cable systems across the country in concert with a hardscrabble rancher-turned-businessman, Bob Magness. Over time, the two men would turn TCI into a telecommunications industry power, the largest cable company in the land, one both admired and feared by industry peers and suppliers.

For the most part, Malone and other industry pioneers, like Comcast co-founder Ralph Roberts, a one-time belt salesman, were committed to the core product: television. But even in the early days, there was a sense that cable systems might be able to do more than just transmit TV channels. In Columbus, Gerry Laybourne's employer, Warner Communications, was attracting attention around a new purpose for its cable lines. Pressing buttons on a custom-built remote control, subscribers could select TV channels, reply to instant polls, order first-run Hollywood movies, and even sign up for music lessons from an instructor who strummed an acoustic guitar at the QUBE studio.

* * *

On a Monday evening in September 1982, some of the cable television industry's top executives gathered in the grand ballroom of the Plaza Hotel near New York's Central Park to hobnob and enjoy "The Cable Follies," a kitschy mashup of songs and stage bits meant to poke fun at their own industry. It was mostly a good-natured roasting of well-known figures like Paul Kagan, an oft-quoted, California-based industry analyst known for his newsletters printed on pastel-colored paper.

The "Follies" were the creation of the New York chapter of a nonprofit industry group, Women in Cable. Ellen Cooper, one of the producers of the debut 1982 show, was a media relations specialist for the cable channel Showtime. Cooper worked alongside dozens of volunteers who brought a polished theatrical flair to the event, raising money to hire a professional stage producer, a New Yorker named Dana Coen, along with a professional music director. "No awards. No speeches. No boredom!" promised a printed brochure that poked fun at the cable industry's penchant for its numerous conventions and get-togethers.

Out of 16 songs presented that night, it was the sixth that stood out. As the lights dimmed, a singer named Laura Wehrsten, by day an office manager for Home Box Office in Manhattan, strode to the microphone. A lone spotlight shone as a plaintive piano chord hushed the crowd. With the room silenced, Wehrsten began a melodic whisper in a sly remake of the song "Tomorrow" from the Broadway hit *Annie*.

"*We'll wire the schools . . .*"

Then she broke for a delicious and extended silence, with everybody in the room understanding exactly where this was going.

"*Tomorrow.*"

It was a revealing parody surrounding a theme everybody knew about but few were willing to talk about openly. For the last several years, companies like TCI, California's Viacom Cable, New York's Warner Amex Cable Communications, and other industry powers were engaged in an all-out war to obtain the legal rights to wire towns and cities across America. The competition for so-called "franchise" territories was fierce. Winning the nod from city councils and mayors usually meant a single company would have the exclusive right to build and operate a potentially lucrative cable television system throughout the city boundaries.

Thus, a robust competition among cable companies to outdo one another with promises for civic elevation ignited in the early 1980s. Wired schools, TV channels devoted to city council meetings, and "institutional" networks that would let government agencies swap information across town were common offerings. In Sacramento, California, during the spring of 1982, the impress-the-cities approach reached a sort of pinnacle, as companies including Warner Amex and the Long Island–based Cablevision Systems Corp. bent over backward to wow the city leadership with promises of the world's most advanced data network—not that anybody fully understood what that meant. Cablevision Systems, established by the industry pioneer and Home Box Office founder Charles Dolan, had promised the city fathers discounted cable TV rates, coupled with a free alphanumeric keyboard that would allow Sacramento citizens to conduct banking transactions, send electronic mail, and more. If that sounds rather like today's World Wide Web, that's because it was. It's just that nobody knew it at the time.

In truth, Cablevision Systems and other companies bidding for attractive cable franchises mostly wanted to extend the physical footprint over which they could offer their core service: TV. Futuristic interactive services could wait, and for sound reasons. Achieving the promised feats of the information age presented a huge engineering challenge—one that would arise repeatedly as the cable industry pursued a role in an interactive communications vision that, for years, was not able to find its way. "I-nets, separate cable networks devoted to municipal use, originally were dangled as bait during the ferocious big-city franchising frenzy of the early 1980s," wrote the *Multichannel News* technical editor Gary Kim in 1992. "At one point, as many as one-hundred I-nets were promised or under construction in major US communities. But few were built, and many others atrophied."

A good share of the blame could be placed on the commonplace amplifiers that gave cable TV its heartbeat. Without them, TV signals would die out over long distances—victims of a phenomenon called attenuation. Special-purpose cable TV amplifiers, brought to market in the early 1950s by Jerrold Electronics of Pennsylvania and other manufacturers, did just that.

Amplifiers were a vital part of cable television's infrastructure, to be sure, but they also injected small amounts of unwanted noise into the signal. The more amplifiers that were used across a branch of cable—a so-called cascade—the worse the signal became. In the 1970s, Ron Hranac, an author and cable industry engineer, worked for a cable company that maintained a cascade of 67 amplifiers populating a long stretch of cable, each one incrementally adding noise to that of its predecessors. Pity the homeowner who lived just beyond amplifier no. 67. "You can imagine the picture quality at the end of the line," Hranac recalled.

Unwelcome noise wasn't the only failing of early cable amplifiers. A related challenge involved the direction in which signals traveled. Almost without exception, cable systems were engineered to transmit signals in one direction only, traveling from a central collection and processing building, called a headend, to customers' homes down the line. There was no means to enable the sort of two-way communication that would be an essential underpinning of interactivity.

Not for lack of trying. The leap from one-way signaling to two-way signaling was a longstanding quest, with scattered experiments tracing back to the early 1970s. The journalist Ralph Lee Smith wrote in the May 1970 issue of *The Nation*, a deep-think monthly magazine, that cable television would serve as the foundation for a new "electronic highway," suggesting that one day "every home and office can contain a communications center of a breadth and flexibility to influence every aspect of private and community life." Smith reprised his arguments in a 1972 book titled *The Wired Nation: Cable TV: The Electronic Communications Highway*. Similarly, in a 1972 technical paper entitled "The Real World of Two-Way," an engineering director for the company TeleCable promised that "most of you here today will be building and operating two-way cable systems very soon, if you are not already doing so."

The projections, however, would prove to be premature. From its late 1940s inception through the 1980s, the cable industry hewed almost entirely to a one-way flow of signals, so much so that it was difficult at the time to see how a dreamed-of era of interactive services over cable lines would ever materialize.

Thus, at the 1982 Follies show, the light-hearted theater performance expressed the reality of the moment: Much of the cable industry's promise to usher in the interactive age was exaggerated. In the original version of the song "Tomorrow," the protagonist Annie sings with infectious bravado about the enduring promise of a new day. "Tomorrow, tomorrow, God bless you, tomorrow," she enthuses. "You're only a day away." On this night, however, as industry big shots laughed along, the Cable Follies laid bare a different interpretation. "Tomorrow, tomorrow, God bless you, tomorrow," sang Laura Wehrsten, "You're *always* a day away."

Still, behind the humor, there was a perceptive quality to the reframed lyrics. Over the coming years, important work would occur—fits and starts, successes and failures—that would eventually make the information highway dream an everyday reality. It would take spectacular technical innovation and billions of dollars in capital, but over time the nation's cable infrastructure really did develop into a bidirectional pathway. The installation of two-way amplifiers, the introduction of fiber-optic lines, the embrace of digital technology, and the invention of the cable modem would converge to make cable systems a powerful, bidirectional communications pipe, transforming the medium into a high-capacity vessel for transmitting digital data streams at stunning speeds.

"Some of the things we sang about were things that actually did come to pass," reflected Ellen Cooper more than 40 years later. "It just took time."

* * *

Although there were limits at the time to how many electronic pathways could course through the copper cable pipeline, a flurry of system upgrades coincident with the creation of new cable television channels had swelled the carrying capacity of modern cable systems from the old-school, 12-channel standard. Around the midway mark of the 1980s, most cable companies were settling around a mid-range, 35-channel model—enough to accommodate popular new channels but not so expensive that it would break the bank. Still, with cable channels seemingly popping up every other month, there was pressure to add even more capacity, which meant spending more money on amplifier upgrades and fancier, higher-performance set-top receivers.

In 1984, having survived the onslaught of terminations and watched as GE's TV development, research, and manufacturing group was disbanded, I moved on to GE's "Comband" project, a scheme for addressing concerns that were beginning to burble up around the subject of how many TV channels a cable system could accommodate. If somebody could invent a way to open up more space—literally create space for two channels where there had been but one—we reasoned there would be a surge of interest. A 35-channel cable system could conceivably become a 70-channel cable system.

Comband made for my own personal classroom, plunging me into the unfamiliar world of cable television. As the leader of the team developing a new set-top receiver for the project, I learned how every cable channel demanded a unique, dedicated pathway to the home, a 6 MHz swath of electromagnetic spectrum. I began to think of cable systems as roadways and television channels as cars: every vehicle—or, in this case, every channel—needed a lane in which to travel.

The Comband project also supplied a lesson that revolved around a newly conceived technology known as digital signal processing (DSP), the transformative construct that turns pictures and sounds from their native "analog" origins into the language of computing—the "digital" part. Once transformed, media and information became liberated from the confines of the past; they can be squished, transported, beamed, shared, and experienced across all manner of devices in myriad forms. Digital signal processing is what liberated music from vinyl records, enables Netflix to stream TV shows, and made it possible for a classroom of fifth graders in Oklahoma to watch in real-time over the Internet as an instructor from Chicago showed them how to dissect a frog.

With a small team of engineers, we worked to leverage newly conceived DSP capabilities in order to create room for two channels in the space of one 6 MHz pathway. It was an idea that would have been inconceivable in the legacy world of analog signals—where the electronic information constituting the image, color, and sound of a television scene is transmitted in its native format, without alteration. But it could be possible in the new language of digital, where we'd transmit signals that had been transformed to the 0s and 1s of digital code. With digital signal processing, we could sample the information once and replicate it in subsequent frames, cutting out lots of duplicative information. It was a lesson I'd tuck away.

Then, a final piece of the puzzle: Our goal was to develop a new integrated circuit, or "chip" in computing shorthand. In this regard, Comband wasn't just my introduction to cable television. It was my introduction to the emerging nerve center of the information age.

The Comband lessons would have a huge influence. I learned how cable television worked. I learned how we could use digital signal processing to squeeze more information into a single channel. And I had my first experience in designing a customized integrated circuit. Somewhere in this mix of cable television lines and digital communications, I knew, was the stuff of something important.

I also learned an important reality about business. Timing was paramount. Being first in a category can leave early pioneers badly bruised. Some trailblazing entrepreneurs were already learning that the hard way.

* * *

At a San Diego office park on a weekday evening in 1980, a roomful of job applicants who had gathered for what they thought was an interview were handed a box of crayons and a sheet of newsprint.

"Now," instructed the young man who had placed the "help wanted" ad and stood at the front of the room, "Draw something. Anything."

Richard Robinson, a local illustrator who had anticipated undergoing a real interview, was deflated. He was growing weary with his job as a graphic designer for the University of California San Diego, and this opportunity sounded promising: working as a graphic artist for a new electronic information service called INDAX. But here, surrounded by fellow aspirants with crayons in their hands, the situation seemed borderline comical.

Robinson scanned the room. Its walls had been decorated with a handful of oversized photograph prints, including an image of satellite receiving dishes nestled against a mountainside. Resigned, Robinson drew a rough sketch of the photograph, walked up to the front of the room, and, like an obedient student in elementary school, handed in his work. Robinson was the first to finish the assignment and eager to get out of there.

Four days later, to Robinson's astonishment, he got a phone call from a man named Dennis Garlington, a manager for the cable company Cox Communications, which operated in San Diego. Garlington was the local leader of the INDAX project—and the fellow who had conducted the unorthodox recruiting gambit. "Your drawing was the best of the bunch, and you were the first to finish it," Garlington told him. "The job is yours if you want it."

Thus began for Robinson a career-long odyssey in the emerging category of "electronic publishing," also called "videotex" or, in early incarnations, "teletext." Broadly, it referred to a grab bag of on-screen information, games, transactional services, and news articles that came to life in the early 1980s. Alongside INDAX, there was Gateway, an interactive information service conceived by the newspaper company Times Mirror Co. and delivered over cable television lines; Viewtron, a like-minded effort from Knight Ridder Newspapers and AT&T that began in Florida; and X*Press Information Services, a cable-to-computer venture backed by the cable company TCI and the publisher McGraw Hill (among a brochure's promises: "Track your stocks. Educate your family. Get instant sports updates. Find the best bargains."). All the newcomers owed at least some of their DNA to General Electric, which had been running a computer time-sharing service since the mid-1960s, later morphing it into the early dial-up online service dubbed GEnie.

The rhetoric behind these concepts certainly ran high. In a 1979 pronouncement, William Ellinghaus, then the president of AT&T, had talked about a grand awakening: "We'll be aiming to introduce a global information market that simply did not exist even a few years ago."

But within a short time, most of the efforts fizzled, leaving what *Vice* magazine writer Cecilia D'Anastasio would later call "a quaint vestige of a 1980s retro-future fantasy."

Viewtron's backers shut down the service in March 1986, joining Gateway, Cox's INDAX, the X*Press service, and NABU Network in closing shop after only a few years of operation. The *Los Angeles Times* journalist James Granelli wrote that executives at Times Mirror "believe that videotex, a computer service providing in-home news, advertising, banking and shopping services, is an idea whose time has not arrived—and may never arrive."

A common problem existed for all these aspirants: Where were people supposed to access this amalgam of fetching content? Certainly not on home computers. By the end of 1982, when Richard Robinson was poring over design issues for the INDAX service, fewer than one of every twenty US homes even had a computer—a chicken-versus-egg problem if there ever was. Some aspirants tried to work around the shortage by sending information to TV sets or, in the case of X*Press, offering to give away a Commodore computer.

Beyond that conundrum, a second issue: how to get to the home. Some of the 1980s teletext players sent their pages over frequencies tucked between frames of TV signals (called the "vertical blanking interval"). Others, in need of two-way communications, relied on telephone lines. But the thin copper wires of the telephone infrastructure were optimized for the relatively low bandwidth requirements of phone calls, not the more demanding needs of multimedia content. Specialized T1 lines that had been developed by Bell Labs provided better performance (a data transmission rate of 1.54 megabits per second) but, at $1,000 or more per month, were priced out of reach of everyday consumers. Integrated Services Digital Network (ISDN) technology that came to the residential telecommunications market later in the 1980s would be cheaper but, at 128 kilobits per second, failed to impress much. Meanwhile, cable TV had its own limitations: The bandwidth was superior, but cable companies could send information only one way.

Despite the hurdles, hope for a multimedia breakthrough lived on. In February 1984, CBS Inc., IBM, and Sears, Roebuck and Co. joined forces to create a service named Trintex (later, sans CBS, renamed as Prodigy). Their service revolved around an amalgamation of news articles, simple video games (like the vexing MadMaze), stock listings, and home banking, as well as a mechanism for writing and receiving text messages that was a precursor to e-mail. A few years later, the contours of the most ambitious information service of them all started to take shape. From the leftovers of a video-game delivery gambit called GameLine emerged an idea for a Prodigy-like service—one that would distinguish itself by emanating a bit more pizazz and a hipper brand identity. GameLine's parent, a company called CVC, consisted of little more than a bit of cash in the bank, an office building, and a handful of employees—one a former Procter & Gamble manager named Steve Case—but CVC was resolved to enter the budding electronic information market. Thus were the makings of the iconic online service that would soon come to dominate the field of play—America Online.

As for the cable industry, skepticism still ruled. TCI, a co-owner of the X*Press service, was convinced it had hit on a breakthrough concept: sending live data streams to computers over cable lines. A senior engineer for TCI, Tom Elliott, would later remark that X*Press "should have been essentially developed into AOL." But it was not to be. "We couldn't engage the industry," Elliott recalled. In the 1980s, it was easier for cable companies to make money the old-fashioned way: by selling subscriptions to popular premium channels like Home Box Office. "We'd go in at the system level and the guys said 'Gee, I run a promo for HBO on the weekend and I'll make more money than I will fiddling around with you guys for the next year,'" Elliott said. As Tom Elliott discovered, not everyone believed what AT&T's Ellinghaus had enthusiastically named a "global information market." In my orbit, one skeptic in particular stood out.

CHAPTER 3

VISION QUEST

Good thoughts, good words, good deeds.

Except for skinned knees and bruised shins—everyday tolls of being a kid—I can't remember having cried before, especially out of sheer emotional anguish. Certainly not because of anything that ever happened in the workplace. And certainly not in the front seat of a Honda Civic on a Thursday afternoon. But here I was, sobbing in my garage.

It was November 1987, the year people were talking about a "harmonic convergence," a global, collective moment of meditation designed to encourage world peace. It was easy to root for this synchronized effort from afar. Who wasn't in favor of global harmony? But the term also resonated with me on a professional level, suggesting a different sort of convergence.

It was the vision of a networked community, a connected city, the empowerment of everyday citizens. The global village made real.

My idea aligned with the tradition of Zoroastrian principles I'd embraced since I was a child. Based on the teachings of the Iranian prophet Zoroaster, one of the world's oldest religions holds that the path to virtue is made up of a sort of holy trinity: good thoughts, good words, good deeds. In my home office, a flag embroidered with golden lettering is draped near a conference table, these three phrases etched in Farsi and in English.

I already knew about one essential ingredient I'd need to make this brand of convergence real. Thanks to my job at GE, I learned about cable television. Then, in the fall of 1986, ingredient no. 2 came to light after I'd accepted a job offer with a company named Proteon, a rising player in the new world of "local area networking"—transmitting computer information around the office at breakneck speed.

These were the two main elements of my vision: local area networking meets metrowide cable television, giving birth to a citywide LAN. The idea of marrying these twin concepts wasn't some hazy ideal. It was absolutely going to happen. I vowed to devote my life to it. But I was about to discover one of the hard realities of life: the inescapable presence of a breed of human known as the nonbeliever.

I had practiced my pitch to Proteon's founder beforehand, homing in on the essential appeal, certain he would see the brilliance: Proteon could distinguish itself by marrying the technologies behind the modern office data-networking environment with the fast-rising medium of residential cable television. We would make fast, "broadband" data connectivity every bit as accessible as a subscription to HBO. Leveraging the existing cable television infrastructure would enable us to forklift the power of high-speed data connectivity from the office to the home at a massive scale.

I scheduled a one-on-one meeting in his office, a 1:15 p.m. start time. But right away things went sour. Minutes into my visit, the meeting devolved into a full-on rant, my host gesturing wildly as he towered over me in an office where a succession of congratulatory plaques hung on the wall.

The yelling by itself was nothing new. Proteon's founder, Howard Salwen, was a dapper-dressed, bow-tie-wearing, Porsche-driving, occasionally moody boss who seemed to distrust many of the people who worked for him. Apparently, that list now included me.

I had joined Proteon as director of hardware engineering, encouraged by my former GE/Comband general manager Francis ("Fran") Scricco, who was Proteon's CEO at the time. The job was a serendipitous trade-off of sorts. I'd teach Proteon what I knew about large-scale design and production of high-quality consumer technology; they'd teach me about data networking. My salary of $60,000 a year was a big leap forward. I was going places.

Or at least I thought I was. But Salwen apparently thought otherwise. "If your stupid idea was going to work," he assured me, "somebody would have done it already." He told me the concept of delivering analog television channels and data simultaneously over the same coaxial cable network was a fantasy. The message was clear. I was wasting his time. I was to return to my office. Stay in my lane.

My "lane" involved optimizing our company's data-networking technology using a networking protocol known as Token Ring—a competitor to an incompatible protocol called Ethernet that would soon become the dominant networking technology. Ethernet's origins traced to 1973 when Robert Metcalfe, a young technologist working at Xerox's legendary Palo Alto Research Center, scratched out a two-page memo adorned with hand-drawn schematics that spelled out the essential logic of his novel networking scheme. Metcalfe's invention gave computers and peripheral devices a way to talk with one another in an orderly fashion at rapid speed over a coaxial network, with mechanisms to help tame contention that could otherwise hobble the entire network. Not long from now, Ethernet would play a big role in my world.

For now, though, Token Ring, which guarded against data collisions by allowing only one computer to send data at any time, was our focus. From our office building in Westborough, Massachusetts, I oversaw two shifts of engineers. One team, made up mostly of newly hired engineers—smart kids just out of college along with developers in their early 30s—clocked in during business hours. The other team showed up around 4 p.m. and would work through much of the night. I'd grab a quick meal in between shifts and stick around for both. In a previous job and nearly ever since, I've learned (possibly unwisely) to forsake sleep in order to get things done.

We were definitely getting things done. We were early to the market with a network router—a specialty that would make companies like the tech giant Cisco enormously successful. It was a heady time, one of those moments when you sense you're on the edge of a big shift. I was excited about taking the next step for our product set, a series of circuit boards tucked into individual computers, along with a networking hub and a data routing card that would make computers, printers, and peripheral devices get along.

At least, that was the goal of our technology. But apparently, we couldn't get along with each other. "Stick to your job!" the founder berated me one final time from behind, as I hurried away. Worst of all was a searing dismissal that now lived on repeat in my brain: "dumb foreigner."

It took 25 minutes to drive to my rented apartment in North Grafton. Leaves were still falling as I navigated my favored route, a countryside road passing through the Tufts University campus. In the garage, I shut off the engine and let my head rest atop the steering wheel of my Honda Civic. And cried.

I made my decision before I exited my car: I would resign. The next morning. A risky move, maybe, but I figured I would manage. I was 30 years old, with most of my career still ahead of me. I was willing to work long hours. I had the makings of a decent résumé as an electrical engineer. At GE, I'd had an early indoctrination into the worlds of the digital integrated circuit and large-volume consumer electronics. I had an undergraduate degree in electrical engineering from West Virginia University, and I was the first engineer to complete GE's rigorous Financial Management Program, the company's internal alternative to an academic MBA. Somebody would hire me. Right?

Except, there was another dimension to my defiance. I was convinced what I had dreamed up could actually happen. Why should we have to confine fast data transmissions to an office? If it made sense for computers within an office to be connected—to share knowledge and information and messages instantly—couldn't the same concept be useful citywide, beyond the confines of cubicles and desks and meeting rooms that made up the modern corporate office?

Good deeds. I could think of few more virtuous than bringing to the world a remedy for problems tied to our seemingly insatiable appetite for fossil fuels. Demand for oil had sparked wars, resulted in polluted skies, caused aggravating traffic jams. I believed there was a better way, and I believed we could move a great deal of information without having to burn fuel and wear down rubber tires. This would be my good deed: connecting people and empowering them with tools to accomplish great things.

Even so, I had to admit our company's founder had a point: Nobody had figured out how to extend the reach of a high-speed office data network beyond the office. To the GE part of my brain, however, the instrument of revolution rose up as an obvious one. Cable television lines already snaked their way across towns and cities in most of the country, built and paid for on the back of cash flow tied to television channels. We could leverage an existing wireline network that provided an enormous first-mover advantage in what promised to become an important new sector. Selling TV channels to America was already generating enormous sums of cash that could be reinvested elsewhere. As *Wall Street Journal* columnist Holman Jenkins would smartly observe more than three decades later, "You might even say ESPN paid for America's broadband rollout."

I was determined to realize the promise of a far-reaching broadband network. I envisioned a radiologist from Boston reviewing an X-ray with a colleague from a rural hospital in Maine. A math genius from San Francisco explaining equations to a student in Arizona whose homework assignment was due the next morning. The landlocked campus transformed into a global university. I thought about family connections: birthday videos of smiling eight-year-olds sent to grandmothers and uncles seconds after the candles were blown out. Workers who could do their jobs without having to drive across town.

Don't get me wrong. I had no illusion that moving data across long distances to massive numbers of users at breakneck speed would improve the lot of humankind overnight. But I thought it might not hurt, either. Besides, what did I have to lose?

* * *

I packed up my office belongings the next day, Friday morning, November 20, 1987. My decision provoked protest from some colleagues and astonishment from the founder and chairman, who had dismissed my idea. It was a stressful moment. With a family to support, I needed income. My younger sister Pamela and my mom, who had been living in Texas as Pamela studied at the University of Houston, were returning to Massachusetts to join me and my father in North Grafton. They accepted the news of my resignation with little overt worry. But I knew they were concerned.

Salvation quickly arrived. With help from a recruitment firm, I found what looked to be a perfect fit: a tech company called Applitek run out of Wakefield, Massachusetts, about an hour's drive from our apartment. I joined the team in February 1988 as vice president of engineering. My annual salary was $80,000. My career was still on the rise. I'd barely missed a beat.

Applitek seemed ideal. The company was working on some interesting technologies related to data networking, where I now had some solid experience. It was established in 1981 by a smart, driven entrepreneur, Ashraf Dahod, a former MITRE engineer who would go on to lead a quartet of successful technology companies (including the mobile data provider Starent Networks, acquired in 2009 by Cisco Systems for $2.9 billion). By 1986, Applitek had found its groove: Annual revenue reached $12 million, the portfolio had ballooned to more than 60 different products, and the company looked to be on its way to becoming one of the big players in local area networking thanks partly to a protocol that appeared to be poised to compete with Ethernet and Token Ring. It looked as if Applitek might one day line up alongside titans like 3Com or IBM.

Then, serendipity. I was excited to learn that in certain locations, a few institutional and government customers were starting to explore the possibility of using coaxial cables to connect remote terminals with a mainframe computer. The instrument behind this achievement was a prototype device whose name was a mashup of the term "network interface" paired with reference to a data transmission rate of 10 megabits per second. It all resonated. With Applitek, I'd managed to land with a company that was at work on at least a crude, early incarnation of my own vision.

It seemed like destiny, or at least the hazy contours of destiny. However, issues around cost, the geographic distance of transmission, and general network performance loomed as major constraints. Deployments of the first-generation product had involved connecting mainframe computers to terminals over small-scale, carefully tended private cable networks. The University of Michigan, for example, had roped together a few buildings using a combination of Applitek's technology and the university's private cabling. Similarly, a large US military supply chain facility, Rock Island Arsenal, had built its own dedicated coaxial cable system and was using it to distribute data from mainframes to terminals that were located in separate buildings. These implementations obviously didn't involve commercial cable systems, the type that were transporting the signals of CNN and HBO to the masses at home. But still, I was energized.

Part of my interest surrounded a breakthrough called Unilink. The proprietary protocol was enshrined in a February 1985 US patent that predated the home computer revolution. Unilink was designed to optimize the way synchronous and asynchronous data terminal traffic traveled over longer distances. Intended as a competitor to Ethernet and Token Ring networks, it addressed their deficiencies regarding scale and breadth. For example, Unilink attacked one of the main hobgoblins of Ethernet: the tendency of network performance to degrade due to increasing collisions of data packets—the building blocks of longer data streams—beyond short lengths of cable. When collisions happen, they result in corrupted packets or data that doesn't make it to its assigned destination on the first try and must be repeated, introducing maddening time delays and inefficiencies on the network. These issues create problems that can be annoying for some applications—email, for example—and fatal for others, such as live phone calls, streaming video, and interactive gaming.

Unilink's remedy revolved around a clever idea. The protocol specified a type of reference signal originating from one of the modems that lived within the network as a means to help each connected modem time its transmission so that packets arrived in orderly succession at the central headend. The result is a reduction in the duration of collisions, making for a faster, more robust network. It had the added advantage of automatically switching as needed from an environment with lots of collisions to one where the more active users would operate in a more forgiving environment. The concept was akin to a highway built with dedicated fast lanes for those who needed them.

I understood the importance immediately, realizing Applitek had come up with a smart way to subdue some of the performance limitations of Ethernet over long distances or Token Ring with increasing numbers of connections. Still, it was not nearly what we needed for the market we were pursuing, namely high-performance, reasonably priced Internet access for every home passed by a cable television network. Unilink provided a way to play traffic cop for multiple connections over metropolitan distances, but it was designed to carry data terminal traffic. It wasn't optimized to carry Ethernet packets that are the foundation of today's connected world. What's more, it was designed to operate over private broadband networks, certainly not over a residential cable television system that posed its own set of obstacles.

But it was a start. Somewhat reflexively, we called our new technology Unilink-II, a hastily conceived naming convention that we would come to regret. The fact is we were too busy designing the protocol to think much about what we would call it. Years later—after we'd sold our company—Dahod would contend in a patent lawsuit that Applitek's Unilink protocol, not our successor solution, supplied the secret sauce that empowered high-speed connectivity for the masses. His legal challenge failed. Our successor protocol used nothing that violated the original Unilink patent; it was invented from the ground up to support the evolving Internet and its myriad of new applications while acknowledging that data terminals had become a thing of the past.

As for the job itself, it seemed to suit me perfectly: I was responsible for overseeing product development and motivating a team of engineers to deliver high-quality products. We worked from a nondescript, ground-level building in Wakefield, Massachusetts, just off the fabled Route 128 tech corridor. Shade trees surrounded the outdoor parking lot, and a garage-style door in the back allowed for the loading and unloading of heavy boxes. It was a typical streetside technology complex with a winding parking lot and lots of open space. About the only nearby novelty could be found behind the building: the remains of an abandoned amusement park called Pleasure Island. During breaks, Applitek employees would sometimes wander around fading remnants like a mock pirate ship, now immobilized in the swampy brush.

Behind the unremarkable façade, though, problems were lurking. What I hadn't examined before taking the job was the company's financial condition—its revenue and cash flow generation. My new employer, I quickly discovered, was in trouble. Dahod had resigned shortly before I joined Applitek under pressure from investors who were growing weary of financial losses. Barely a month into my job, problems that threatened the health of the entire company had materialized.

I was shocked, for instance, to learn that a supposedly sophisticated network management system—a self-diagnosing technology that monitored network health and scoured for possible problems—was highly exaggerated. The data showing up on the screen when representatives demonstrated the technology consisted of made-up metrics typed into a word processing application. Vaporware. Other prototypes were hurriedly engineered and shipped as if they were fully productized. I worried that Applitek's sales representatives had been selling "finished" technologies before our engineering crew had fully completed their work. In addition, I began to have serious qualms about the breadth of our product line in general, realizing we had veered off in dozens of directions as we customized solutions narrowly to suit individual clients but neglected any real sense of duty to the profitability of the company or the revenue contributions of individual solutions. It's easy to say "yes" to a client who wants a specialized solution; nobody wants to disappoint their customer. But acceding to every request creates an explosion of problems down the road: too many narrowly focused products sapping a company's resources, draining money, eroding internal confidence, and causing problems.

The fraying was showing up in some caustic feedback from clients. A memo from a colleague listed dozens of incidents being encountered by our customers. One user was "unable to disable ports because of a config change." Another was experiencing "erroneous filtering of physically addressed packets." An Emory University customer was "very concerned with performance on his network." Nynex, the big phone company, was "still waiting for (a) protocol analyzer." At the Wright Patterson Air Force Base in Ohio, the network was "doing strange things," per the memo. "Massive collisions and missed packets" were common. Another big customer, the telecom company GTE, was repeatedly experiencing failures tied to manufacturing defects. On a daily basis, we were slammed with technical issues.

A related concern involved employees of the company. Early during my tenure, they were exiting fast. I buttonholed a few expats on their way out, concerned that it was my hiring as the lead technology executive that had steered them away. Not so, they told me. They'd seen writing on the wall well before. I was advised not to worry, at least not until I'd been on the job for 60 days. If people were still departing at that point, then I should be concerned.

Sure enough, a key Applitek inventor and one of the original Unilink architects, Chris Grobicki, pulled me aside on day 61 of my tenure to tell me he was resigning. Chris was exhausted from six years of elongated workdays and workweeks and had grown disenchanted with Applitek's recent pivot away from the Unilink protocol. "Please don't take it personally, Rouzbeh," he told me in an amiable, low-key style I would come to prize. "But I don't think this company has a lot of value left for me." I was disappointed. These people were outstanding at what they did, invaluable engineers and managers with a spirited, engaged approach to their work.

Dahod's successor as president and CEO was an amiable golf enthusiast and ex–Data General Corp. executive named Larry Holswade. He wasn't a techie, like Dahod, but that was fine with me. As vice president of engineering, I welcomed the chance to have a wide latitude in managing the technology side of our business. Except now, the talent drain meant that my duties were piling on.

Besides engineering, I was tasked with managing customer service, sales, and marketing—fields where I had zero practical expertise. I was relying purely on instinct and common sense as I flitted from one fire to another. In the space of a half hour, I'd pivot from a call to soothe a distressed customer to a demonstration of technology to a review of that week's product improvement efforts. I wasn't the CEO, but it sure felt as if I was functioning as one.

For his part, Holswade remained calm. He listened patiently to my list of concerns and my ideas for solving them. Even though I was still bruised from my confrontation at Proteon, I told him about my grand idea: to transform Applitek into a leading force behind a telecommunications revolution that would marry fast data networking with the breadth and presence of residential cable television lines. It was the first time I shared a name that had been percolating in the back of my mind. We would call the endeavor LANcity, expressing a concept for making a local area network operate across an entire city. His reply was vintage Larry: "Come up with a budget and a staffing plan, Rouzbeh." He'd hear me out upon returning from the golf course.

Worry was setting in. The company I'd seen as a vehicle for creating a connected world appeared to have multiple advancing illnesses. If Applitek was going to take us to the promised land, it was apparent it would have to do so with severely limited resources and an investor group that was tiring of ongoing struggles after almost seven years of effort.

Still, I kept my faith in the bigger ambition, holding tight to my vision of the connected community. But it wasn't going to happen simply because I wished it to be. There had to be a launching point. During evenings, after a momentary easing of the day's turmoil, I sat at my desk and began to trace out the contours of an actual specification. It would be the blueprint for a technology I believed could change the world. With the office hubbub quieted, I drew a triangle on a notepad and scratched out the core requirements for a new device I would call the "cable modem." (The "mo" terminology is short for "modulation," or the attachment of data to a slice of RF spectrum; the "dem" addresses demodulation, or the extraction of the attached data from its assigned carrier wave.)

Bucket no. 1 for my cable modem called for a new integrated circuit that would support data networking and digital processing functions. My current concern—which was a major worry—was that Applitek's first-generation product set handled modulation the old-fashioned way, relying on a succession of analog physical components. Convincing hundreds of these piece parts to work in harmony within the pathways of a circuit board required a vexing interplay of physical and mechanical adjustments that taxed even the most talented of electrical engineers. Paul Nikolich, an Applitek developer who surely met that description, calculated that there were more than 40 separate, painstaking alignments needed to make it all work. At the time, it took an entire workday to get just one of the units ready. Applitek, recognizing the challenge here, had considered developing its own customized, high-density integrated circuit but ultimately determined that there wasn't enough sales volume forecast for the product line to support such an ambitious, costly development effort. But the thinking, at least, was directionally correct. I knew that to attack the problems of size, cost, and complexity, we'd need to exchange our interior intelligence from the realm of the analog circuit board to the alchemy of a digital integrated circuit. The problem was that nobody had built a chip of this magnitude in the data networking field. I wasn't certain anyone could.

Bucket no. 2 involved a media access control protocol (MAC for short): the set of instructions that tells bits of data how to find their way around the network, sling-shotting from sender to receiver. The MAC protocol we would require for a high-speed modem would need to be reengineered entirely. I recognized early on that the original Unilink protocol was worlds removed from being able to accommodate massive numbers of simultaneous users receiving and transmitting data streams across residential cable television systems.

Then, bucket no. 3: fundamental considerations around size and cost. Even with a powerful new integrated circuit and a powerful new MAC protocol, nothing would work without a wholesale reengineering of the surrounding hardware. Applitek's product weighed nearly 80 pounds, relied on a succession of 12-inch-by-12-inch circuit boards decorated with hundreds of components, and was roughly the size of a small dormitory room mini-refrigerator. If you wanted one, you'd need to write Applitek a check for nearly $18,000—not exactly a price point suited for mass deployment in the residential market.

Over the next few months, I shared the ideas privately with a small, core team: a mix of my top developers and a few outside advisors. A few of my colleagues stayed late at the office, hearing me out and encouraging me to keep the faith. The specification we were devising described the main ingredients of an information economy that would begin to take flight in the mid-1990s. For now, though, we had identified our buckets. And they were empty.

* * *

Trouble had been brewing in Washington D.C. around the time I'd joined Applitek, with the nation reeling from disclosures tied to a secretive scheme conceived by Reagan administration operatives, reportedly without the knowledge of the president. The scandal involved the illicit sale of weapons to Iran, with proceeds secretly funneled to a band of Nicaraguan rebels. Headlines called it the "Iran-Contra Affair."

Like almost everyone, I recoiled at the brazenness of the scheme. But the semantics left an interesting imprint. "Contra" meant rebellious, contrary, against the grain. My bucket list was surely that. The Iran-Contra controversy was receding after several years of scrutiny, but it had widened my vocabulary. I started calling what I was about to try to pull off our version of a "Contra plan."

The Contra plan started in a living room in September 1988. I wanted us to get out of the office, partly because our conference room was small and partly because the surrounding vibe was increasingly depressing. So I gathered our core team in Reading, Massachusetts, at the colonial-style home of Mary Lautman, a smart software developer who'd stuck around, maintaining faith as the company reeled toward what was beginning to look like an unstoppable downward spiral.

We sat on the floor in front of a wooden easel I had carted over. On oversized sheets of white paper, I sketched out a list of all of our products, tracing horizontal lines in soft blue ink and then intersecting them with vertical columns to produce a matrix of blank boxes. Across the top, I listed six categories gauging the performance and promise of each product: "revenue production" and "important new platform" were two of the criteria. Underneath, and running down the remainder of the sheet, were the names of more than five dozen distinct products—the entire Applitek catalog.

The environment in Mary's living room was cozy enough. But this was a sobering conversation. Unshackled from the formalities of the office, my team let loose. Almost every product we shipped seemed flawed in some important way. Too little revenue, too few customers, too much engineering development work still needed. In a memo, a colleague, Steven Dunstan, pointed out that some of our products hadn't been modified in five years—a lifetime in the fast-moving data networking arena—and were 30 to 50 percent more expensive than what clients would now pay for a modernized successor.

Product after product, we identified shortcomings. The column at the far right of my matrix was labeled "keep/kill." Almost everything fell into the latter category; we were willfully murdering most of the product portfolio. An alternative path would have been to neatly trim the list here and there, pruning in the same way I'd watched my father neatly paring back branches of fruit trees in our backyard. Careful, methodical, mannerly. I admired his patience and reverence for his craft. It was his happy place. Except, it wouldn't work for this garden called Applitek. We didn't need to prune trees. We needed to uproot them.

Making that happen would require a serious balancing. On the one hand, we'd have to continue supporting clients who used our (admittedly) dated hardware in order to preserve the monthly service revenues that were keeping us alive. At the same time, however, we'd stop putting money and staff time into technologies and systems that had no future. Instead, we'd hitch our company's future to a single system.

The lone survivor would be our first-generation modem—by now, I had started calling it a "cable modem." Otherwise known as the "NI-10E," it was the one product in Applitek's arsenal that could conceivably address a large-scale, emerging market. And it was the only product that aligned with my vision of turning residential cable television into the world's biggest data network. We'd be pursuing parallel objectives: Keep the company alive, on one hand, and invent the future, on another. Neither would be easy, even by itself. Pulling off both simultaneously would be a long shot.

As for the mainstay product itself, it required two people to carry it. It was gawky looking, an upended rectangle with slanted legs jutting underneath to keep it from tipping over. On a good day, we could get a single unit built and out the door. The 1989 movie *Honey, I Shrunk the Kids*, would provide an apt metaphor, with me substituting the word "dinosaur" for "kids." The NI-10E was our dinosaur: big and expensive. Still, it reliably ran the aging Unilink protocol that orchestrated the distribution of data packets from point A to point B. It could transmit data at a speed of nearly 10 megabits per second, traveling over a 6 MHz swath of radio spectrum. It was performing adequately for customers, including the midwestern US military logistics facility Rock Island Arsenal, the University of Michigan, Vandenberg Air Force Base, Emory University in Georgia, and a few international deployments. Even so, I recognized the world was soon going to pass us by. Advancing technology would leapfrog a product that was meant to connect data terminals to mainframe computers, not household computers. Recognizing that Ethernet connectivity was becoming more prevalent, we were able to extend Unilink's capabilities to interconnect disparate Ethernet networks, but only as a stopgap until we had something better.

I'd done my best in the living room that evening to evangelize the vision, to lay out a path for making something important happen. We all wanted to achieve something, to have a purpose in our careers. But we knew our situation was tenuous. We had zero access to additional funding. What we needed was what lots of bootstrapping tech companies need: one very large, supportive partner that not only had financial resources but faith in our vision.

Identifying that company would be a long shot. But in the background, there was a faint tease of light. A few months earlier, I presented ideas about data networking over residential cable television systems at an industry conference in Boston. Stepping down from the stage, I was greeted by an enthusiastic pair of men who told me they worked for Digital Equipment Corp., the revered computer and technology company headquartered in Maynard, Massachusetts, not far from our office. They mentioned they were also starting to explore the idea of a new sort of data-networking scheme, theorizing that it might be possible to take the magic of the LAN and apply it beyond the office. They, too, had wondered about those long expanses of coaxial cables coursing through towns and cities. I couldn't believe what I was hearing. Digital Equipment Corp. (or DEC) was one of the revered overlords of the LAN category, credited as much as any other company with introducing Ethernet technology widely across the enterprise domain.

As fast as I could, I spat out details about the work my company was doing and how it could be the foundation for a new sort of cable modem that connected to cable television systems, an idea that some people found preposterous but I found irresistible. I told them about the two essential standards behind my vision: Ethernet and the 6 MHz television channel. They seemed genuinely intrigued, understanding these were the stars that would need to align to achieve the vision I'd described. We exchanged business cards, shook hands, and pledged to stay in touch. I memorized their names: Jim Albrycht and John Kaufmann.

In the world of printed circuit boards, the term "signal trace" describes a flat, narrow copper pathway uniting disparate components. These connections are tiny, typically measured in "mils" or thousandths of an inch. But the size isn't what matters. It's their role as junctions, uniting disparate sources of energy, which makes things happen. I didn't realize it at the time, as I hustled out of the conference room, but my own signal trace had just been set into place.

* * *

It's close to 2 a.m., early December morning, 1988. I'm driving north on California's Highway 1, hugging tight to the coastline. I'm terrified. I had no choice but to hit the road to convince our customers to stay with us. We were clinging to life. Applitek's annual revenue, once more than $12 million, had collapsed to less than $2 million and was shrinking rapidly. The company was looking more and more like a train wreck: customer problems were rising, complaints were everywhere, and employees were resigning. Eroding from the inside out.

Larry Holswade was gone by now, having decided Applitek was a career dead end. Our new temporary CEO, a former Honeywell developer named Nick Papantonis, was mostly a stand-in, charged with keeping the lights on while our investors figured out what was next. The employees who remained on board, keeping the faith despite all evidence to the contrary, had made a bet based on a parallel reckoning: First, that I might possibly know what I was doing. Second, that what I was doing might actually matter.

My schedule had been a blur: I'd flown that morning from Boston to Chicago to visit the systems integrator who oversaw our Rock Island Arsenal deployment. From Chicago, I flew to Los Angeles International Airport. From LAX, I collected my rental car and began what would become a perilous drive, plowing face-first into an unyielding fog in the deep of the night, on the way to Vandenberg Air Force Base.

I now understand what people mean when they say a fog can be as thick as pea soup. If there was a car in front of me, I couldn't see it; the atmosphere was billowy cotton, obscuring even the barest hint of a red taillight that, for all I knew, loomed only a few feet ahead. All I could do was slow to a crawl, try to make out the faint contours of the road's curves, and hope. How would I even know when I've reached my exit? The only guide I had was a folded map from the Hertz car rental office that rested on the passenger seat. I could pull over, but I'd be at risk of being slammed from behind by an oncoming driver. I kept my hands gripped tight to the steering wheel, and I negotiated a private bargain: I let God drive the car.

God proved to have solid navigation skills.

Somehow, as if by decree of a forgiving universe, there was a brief window of clarity. I could see the green highway sign telling me my exit was next. Thank heavens. I had never been so relieved. I exited the foggy highway, found my way to the little roadside motel, and steered the car inside the painted outlines of a parking spot, elated to turn off the engine.

It was now past 3 a.m. I had an 8 a.m. meeting the next morning, one of dozens of face-to-face conversations—confrontations, in many cases—with clients whose ongoing service payments provided the narrow strand that kept us alive. We had completed our assessment of our technology as part of our Contra plan, and it was time to meet with our customers. After an abbreviated sleep—I was still wired from the scary drive—I showered, dressed, met with my Vandenberg clients, drove back to LAX, and boarded a plane for Minneapolis.

My four-week road trip was an attempt at contrition. I was doing something we had not done before with any degree of authenticity: listening to our customers and trying to understand and address their problems. I was convinced that, otherwise, we'd lose our financial lifeline: the recurring service revenue that kept us afloat. They were our only practical hope.

As expected, the feedback was mixed. The Vandenberg people were perfectly polite, but a stop I'd made earlier at the University of Michigan was tense. There, the head of the Information Technology department berated me for an hour over poor product performance. We met in a basement computer room, where racks of blinking boxes produced constant noise from whirring fans, causing him to raise his voice even more. Halfway through, he was yelling at a near-comical pitch. Still, I listened patiently. By the end of our hour, he calmed down.

So did others. I realized that the more customers talked, the steadier they became. Here, finally, was someone with technical knowledge and managerial authority who seemed to be willing to hear out their complaints. In almost every case, I came in as a hostile foe but left as a friend. I opened every conversation with a variation on a simple question: "What do you guys think?" As the conversations flowed, I learned an important lesson: Never sell a customer something you don't have. Never pretend. Never fake it. Let your technology—not your brochures or glib pronouncements—do your talking.

Although our relationships warmed, the road trip produced bad news: More than half of our customers were divorcing us, fed up with years of lackluster support and the absence of any genuine sense of responsibility. Fair enough. At least they were honest. I came back to Wakefield with a list of just six customers we could count on to continue the business relationship. It wasn't encouraging. But at least I now knew what our service revenue projections were going to look like, which was to say, not great.

* * *

The town of Stow, Massachusetts, is nestled into a gentle hillside in Middlesex County, 25 miles west of Boston, a postcard-worthy New England community where motorists drive slowly across rustic county roads and apple orchards reliably flourish every autumn. Stow's origins trace to 1660, when, per local lore, a settler named Matthew Boon traded a jackknife to leaders of the Indigenous Nipmuc tribe for rights to farm a patch of land.

Not listed in the town's official historical record is another "first." Stow is a historic milestone marker on the road to the information highway. Thanks to a great deal of persistence. And to a man named Nick Signore.

A gregarious, can-do optimist, Nick had worked early in his career as an auto mechanic, finessing front-end alignments and wrenching torsion bars before deciding to abandon the garage and instead do what many young professionals were doing: get a degree in information technology. He landed a job with MITRE Corp., a US Air Force contractor based in Bedford, Massachusetts, and later joined Applitek.

In the fall of 1989, I sent Nick to Stow to conduct a reality check essential to the vision of the connected city. Nick would be paired with a young technician who worked for a small cable company, Nashoba Valley Cable TV. The company's management had agreed to allow our team to rig up a first-ever test of our cable modem technology over a live, real-world, residential cable system.

This would be a big moment. Other clients—universities, government agencies, military contractors—ran their data networks on the backs of private coaxial cable systems. They were specially built, dedicated, and privately managed facilities—worlds away from the types of residential cable systems that laced their way across most of the nation. Instead, these private systems, operating at deployments like Ohio's Wright-Patterson Air Force Base, were ideal for data transmission from an engineering perspective. There was no contention for bandwidth because there were no TV channels to worry about. It was purely digital data coursing through the pipe. What's more, the amplifiers powering these private systems were top-of-the-line, two-way devices—precursors of what cable television technology might one day become. The systems were like the dedicated fast lanes of a highway, available only to a select group of motorists.

In Stow, however, the cable lines that would be used for our data streams were the same lines that whisked the signals of CNN and ESPN to users' TV sets. Any disruptions we caused could affect paying subscribers, who relied on Nashoba Valley Cable to get their HBO and their MTV.

We weren't naive. We knew there would be technical problems. But that was the point. In order to see our vision through, we'd need to identify obstacles and figure out a way to overcome them. And so, Nashoba Valley Cable set the stage: It would be the launchpad for the global connected village.

It was important that I demonstrate to my new acquaintances at DEC, and my own crew as well, that we could get it all to work. On paper, the task was straightforward. We would connect the home of my new friend Jim Albrycht, a two-story house on Peabody St., to John Kaufmann's residence on Packard Road, a cul-de-sac about two miles away. For months now, the three of us had been meeting at least weekly, sharing updates on our data networking projects and technology development efforts and cheering it all on.

Albrycht placed the call to Nashoba Valley Cable, introducing himself as a DEC engineer and a local customer. The "ask" was for a dedicated slice of frequency. We explained to the system's general manager that we wanted to demonstrate how our technology could work within the established contours of cable television, riding over a 6 MHz cable slot. Luckily, there was just the thing: Nashoba Valley Cable had reserved a channel toward the higher end of its spectrum for the city's own use. The slot had been set aside for a community channel that could be used to televise town government meetings and other "public access" programming.

Beyond having a spare channel, the cable system had been outfitted with two-way capability that one day might be used to support bidirectional interactive applications involving local schools. That was an unexpected bonus. We'd presumed we'd have to rig up two-way amplifiers ourselves. Even so, we still had to prove it would work.

We got to work. At the Nashoba Valley Cable central office (the headend in cable vernacular), we connected a translator: a device that converted the outgoing data signal from Albrycht's house to a downstream frequency that would find its way to the Kaufmann house. It was a bit unsettling to see our hulking NI-10E sitting on the floor in Kaufmann's living room, running up the electrical meter and billowing noise from its multiple fans.

Even more unsettling, it wasn't working. On the monitor connected to a DEC workstation and in turn, to the modem, nothing was happening. No data. No action. Nothing.

I was dumbfounded. We had a working translator at the cable company's headend, two working modems at either end of a two-way span of cable, and a perfectly serviceable workstation connected with a standard Ethernet wire. Over this connection the Unilink protocol was playing its familiar role of data traffic cop, as it had done elsewhere.

We knew this combination of technologies worked. It worked at Emory University. It worked at the University of Michigan. It worked at Vandenberg Air Force Base and at Rock Island Arsenal. But it wasn't working in Stow, Massachusetts, over a real-world, residential cable system.

I was perplexed but also determined to understand what went wrong. The following week, I asked Nick and our engineers to return to the scene of the crime: the stretch of cable that ran from our headend translator to Kaufmann's living room. I armed Nick with tools that would sniff for trouble: a portable generator to deliver 110 volts of electricity to power our modem and test equipment, a spectrum analyzer to visualize the signal, and a signal level meter that would tell Nick how strong our signal was at any measured interval.

Nick and the Nashoba Valley Cable technician drove slowly along the streets of Stow. Nick rode shotgun in the passenger seat, squinting at a curled blueprint. A blue triangle on the map indicated where they would find overhead line amplifiers. Once they'd spotted an amplifier attached to an overhead line, they pulled over. With Nick guiding from the sidewalk below, our technician elevated slowly to the assigned location in a single-person bucket attached to a hydraulic-powered utility truck. They ran temporary feeder cables from the amplifiers to the modem powered by our portable generator.

It didn't take long to identify our culprit. The signal heading toward Kaufmann's house was mired inside the spectral equivalent of a municipal trash site. Noisy, messy, laden with electrical interference. The signal we thought could run over a couple of miles couldn't get past the second amplifier in the same neighborhood. Nick watched in dismay from the sidewalk as his signal level meter told the sorry story.

It was immediately apparent that our technology couldn't deal with the imperfections of a real-world residential cable system. This was no minor engineering problem. Nashoba Valley Cable TV was small, but it was a microcosm of the entire category. If our technology wouldn't work across two amplifiers in Stow, Massachusetts, it wasn't likely to work in any residential cable television environment, certainly not over extended distances. I knew from my work at GE that cable television systems in the United States were almost uniformly alike. Amplified cascades of coaxial cable connected millions of homes. If our signal couldn't traverse two amplifiers, how was it supposed to survive over more than 50 of the devices, stretching across entire neighborhoods?

In the pastoral town of Stow, Massachusetts, I began to realize what we were going to be up against.

CHAPTER 4

IT'S A DEAL

I had a secret superpower: nothing to lose.

I had shared my updated revenue projections stemming from candid customer conversations with our board of directors: representatives of six companies led by the venture arm of the Boston money management firm Fidelity Investments. The board members were happy to see some semblance of predictability, but they remained nervous. None could see the promise of a return on investment. From their vantage point, backing this company looked like a mistake.

I disagreed. I was convinced that somewhere between Applitek's aging technology and our specification for a new, second-generation cable modem was a formula for renewal. As the saying goes, there was a pony in here somewhere, and I was willing to bet my career on it. In August 1989, I proposed a remedy during a meeting at our Wakefield office: a management buyout. My team would take over ownership, paying the investors back not with a lump sum—we had nothing in the bank—but based on an earn-out schedule that would deliver a series of predictable quarterly payments.

The board members politely heard me out. But it was obvious they saw little hope for what I was promising. They'd concluded there was no way for me to eke out enough profit to generate a payback. They also recognized a second reality but were kind enough not to articulate it out loud: I was a tech guy. An engineer. I had zero experience running a new start-up company.

Instead, the board decided to hire a new CEO, the company's fourth, in a last-ditch effort to salvage Applitek. Asbjorn Sorhaug was a boisterous, likable Norwegian, a former military pilot who built small airplanes and exuded a fresh, can-do spirit. He would show up in the parking lot now and then with a car trunk full of fresh Norwegian salmon on ice, offering the fish at a discount to his new colleagues. Sorhaug convinced the board, having turned down my buyout offer, to undertake what seemed to be the only logical course of action: sell the company for whatever they could.

And so they did. On January 15, 1990, the industry trade magazine *Network World* published a seven-paragraph story about Applitek's acquisition by an Australian company, Computer Protocol Corp., a company described by the magazine as a provider of "multiprotocol communications servers." In the article, our new CEO praised the combination, saying it "takes advantage of complementary products and markets." I was quoted, too, enthusing about presumed synergies.

The press coverage was flattering. But I worried about the stated kinship between us and Computer Protocol. It seemed to me we were playing in different sandboxes. I wanted to convert Applitek's technology into something extraordinary, the centerpiece of a new data revolution. Computer Protocol wanted to expand its hardware business beyond its core markets of Australia, Southeast Asia, and Europe. The deal was mostly a way to penetrate the world's largest high-tech market: the United States.

Things quickly went awry. Looking to get Applitek off their books and at least capture a tax write-off, our investors made questionable concessions, the main one having to do with cash—or lack thereof. Rather than pay the promised sum at the deal's closing date, our new owner offered a promissory note—a written pledge to pay our investors a few months later. Now, any cash we produced would flow to their coffers, in exchange for an IOU.

As I feared, the "owe" part of the acronym proved to be problematic. The promised payment, originally scheduled to be made within 90 days, never materialized. Instead, our new ownership asked for an extension. Months after the transaction was announced, we hadn't seen a dime.

And yet, our expenses were mounting. Two sales representatives who were part of the deal traveled extensively, racking up big credit card charges we could scarcely afford. As head of sales and marketing (one of several new roles I was saddled with), I worried about our rising American Express tab and our dwindling bank account. By now, Sorhaug had exited, and my dream was fading before my eyes. By May 1990, we were on life support. We were close to losing everything. Everybody knew it—everybody, including a polite, street-smart executive named William Elfers.

Elfers—by now "Bill" to me—was a Fulbright Scholar who worked on Wall Street after graduating from Harvard Business School. He was the point man for our investor group and the manager of Fidelity's venture capital arm.

Fidelity Capital was Applitek's biggest investor, having sunk money into the company on the faith that it was poised to do big things in data-networking technology. If anybody was going to take the heat for our failure, it would be Bill, an exceptionally calm, logical individual I had come to appreciate and admire. He approached me from time to time for help in understanding the real story behind the technology and product development work we were doing, and I was always happy to guide him through our engineering work. I respected his genuine intellectual curiosity about our craft—a trait not always in large supply from the money men.

I was at a crossroads. My team and my vision were all I had. I was willing to work overtime to develop new technology. But I had zero confidence in our new owners. I called Elfers, explained my concerns, and asked for another face-to-face meeting with the board. I was certain that unless something dramatic happened, we would fail, and soon. We were barely two weeks away from missing payroll unless a much-needed check arrived in the mail within a few days. Again, I re-stated my offer: I would pay back the investors on a quarterly schedule, collecting every available dollar of spare cash we took in and sending it to them. Based on predicted payments from our customers and the remedy of suspending my own salary to even out the cash flow, I figured I could pay them $40,000 every three months for four years. The total sum would be $640,000, all of it derived from existing service revenue streams. Nobody was going to get rich, but it was an alternative to our investors writing off their losses.

I made sure the deal terms were crystal clear. The Applitek investors would have no role in our day-to-day operations or governance. They would purely be silent investors. I would run the company independently, operating as a sort of commonwealth. I would lead the charge, relying heavily on input and ideas from a core team of 13 key employees, advisors, and friends I hoped to recruit.

I showed up at Fidelity's downtown Boston building to repeat my pitch to the board. What I got was mostly silence. Elfers was kind enough to join me on the elevator ride down to the first floor. Rounding the corner toward a coffee shop, I pressed my case once again: Why shouldn't I take over what was left of the company? It was either that or a near-certain bankruptcy filing, one that might make our investors look foolish. I knew this was a long shot. I had zero practical executive management experience. Although I had earned a degree from GE's vaunted Financial Management Program, I had never been a CEO and had never had profit-and-loss responsibility for an entire business. But I knew our customers, and I knew my team. We had our Contra plan. I had a hunch that DEC would end up in our corner. I had an unwavering conviction that despite the odds, we could make an impact.

We stopped in front of the coffee shop. "Rouzbeh," Elfers said, "do you have any idea what you're doing?" A fair question. But still, the only choice for the investors was to take a flyer on an unproven businessperson who professed to be willing to turn the thing around, or record their losses and suffer an embarrassing black eye within the clubby Boston investment circuit. Elfers, Fidelity, and the handful of partners had little to lose. And to his credit, Elfers knew he could do what I could not: sell the idea to the board.

And so, we had a deal.

As of June 19, 1990, the Australians were gone, and I was the new CEO. The board had agreed without protest to my terms. I'd write a check for $40,000 every three months for the next four years, recording the payments in a leather check register that would become my ever-present office companion. When that last payment had been made, I'd be in the clear. Meantime, there was pressure baked into our agreement: If I missed a payment, the investors would get the company back, along with the deed to my house, which I'd pledged as collateral.

The payback number was calculated based on an exacting, down-to-the-dollar accounting of our business. In theory, I could make our payroll, buy materials, pay our office rent, placate customers, run our business, and pursue my vision while scraping together the required payment every 90 days.

I would need to suspend my paycheck as well as my younger sister Pamela's paycheck several times over the next four years, but I held fast to the end of the deal. I had 13 employees, an aging product, half a dozen remaining and loyal customers, a rented office in Andover, Massachusetts—we had moved to smaller quarters to save operating expenses—and not much else.

I also had a new name. I had come up with the moniker LANcity in 1988 to express the idea that local area networking technology, already revolutionizing corporations and the workplace, could be deployed city-wide. I incorporated LANcity around the time I made the first payment to our backers. My colleague Les Borden drew up a logo, a graphic of a cityscape nested atop a length of coaxial cable.

LANcity was officially born. It would sink, languish, or prosper based on a matrix of decisions I would make. Implementing our product set, making payroll, keeping customers happy, and sending checks every three months to our investor group were my focal points. Late nights and working weekends would continue to be the rule. Vacations were inconceivable. Welcome to being a CEO. And a janitor at the same time.

On the plus side, I had a secret superpower: nothing to lose. It was possible we might end up failing. But we held firm to the vision: data coursing across the residential cable TV pipe at high speed, over long distances, connecting everyone.

At least there was one bright spot. We had tweaked and engineered our mainstay product to the point where the platform had stabilized and was reliably working—unless, that is, the Mississippi River flooded.

* * *

"You get down here and fix this thing in six hours or I'll have you arrested!" he yelled. The voice on the end of the phone belonged to Bud Walters, a civilian who ran the data network powering operations at Rock Island Arsenal, a US Army logistics and weapons management facility located near Davenport, Iowa. Walters reported to a much-decorated general who ran the island facility. He and I had hit it off, engineer to engineer, when I'd visited him on-site in 1988. But now, he was livid.

If there was ever a facility that demanded failsafe information capability, Rock Island Arsenal would top the list. When Applitek first began operating there, the arsenal was focused on the Defense Standard Ammunition Control System, a US Army program to supply what officers called "war-stopping munitions." The arsenal lords over a 946-acre slice of land atop the mighty Mississippi River (between Davenport on the northeast side and the Illinois town of Rock Island below). Army engineers built levees to guard against washouts and erected imposing limestone walls that dared storms to try. The effort was deemed to be worth it: The unique geographic setting—a self-contained island nestled atop a vital shipping route—made Rock Island strategically attractive.

But now, in the fall of 1990, rain kept coming. Sheets of it, engulfing Rock Island and knocking out a key part of the facility's nerve center: the data network used to shuttle mission-critical information from one part of the facility to another. The same data network that depended on our product line. Lately, the network had encountered rising demand as users embraced new applications like electronic messaging. Dick McGarry, a LANcity field engineer who worked on-site, remembered the excited buzz surrounding the arrival of close to 1,400 Zenith personal computers. Until then, users had to reserve time at one of 14 workstations in order to send a text message to any of the terminals placed throughout the arsenal. Now, with the arrival of new personal computers, hundreds of simultaneous users could zip written messages to any other connected user by typing an eight-character ID. The newfound ability was a revelation.

But no messages were zipping around at this very moment. Nothing was. The flooding had breached the protective rubber shields surrounding underground copper cables. The network was down. I was on the hook here because I was, in military parlance, the commanding officer of the company that supplied the gear powering Rock Island Arsenal's data network. Although we were contracted through a third-party systems integrator, the gentleman on the line apparently wanted to go directly to the presumed source of the problem: me.

We couldn't afford to lose Rock Island Arsenal. The Army facility was by far our biggest customer. It's why Dick McGarry was on the scene full time. Dick, who lived in Davenport and had once worked repairing Apple computers at a local store, had impressed the manager of the systems integration firm that had responsibility for the Rock Island IT infrastructure. When that company was sold, Dick came aboard as Applitek's point man on the scene.

Besides its importance from a day-to-day business standpoint, Rock Island Arsenal was an ideal laboratory for understanding the future. With more than 5,000 users and an escalating cascade of data transmissions, I was certain the facility ranked as the largest cable-based data network in the world. Other LANcity installations served only a handful of users. Rock Island Arsenal was sprawling by comparison. Because of the network's breadth, we were ahead of anybody else in confronting and dealing with a wide range of issues that later would occupy the entire cable industry: how to calibrate signal levels across a wide user base; how to shape and manage traffic at scale; how to deal with naming conventions tied to TCP/IP, the language of an emerging Internet. We learned how demand levels would vary: For instance, we came to expect a surge of activity early in the morning, with a sudden decline around 9:15 a.m.—coffee break.

We also learned about some of the tricky physical considerations and constraints that would be involved in the outside world. Many arsenal buildings were constructed from locally quarried limestone, their walls several feet thick. The only way to get a signal inside a building was to drill through the stone and thread a length of coaxial cable inside it. Dick Beard, a telecommunications construction contractor who had been tapped to build the Rock Island cable system, credited Rock Island Arsenal with introducing his technicians to the wonders of the supersized, water-cooled power drill. Thanks to their labor, data streams were now coursing over these cables at a speed of close to 10 megabits per second as they shot through the walls.

Beard's company had been commissioned by an Army general contractor to build a robust, high-capacity cable system: all underground, with an impressive 750 MHz of capacity and modern, two-way amplifiers throughout. It was a microcosm of the sort of state-of-the-art cable system that would later emerge across the United States.

Here, our technology could shine, drawing from its interior smorgasbord of hardware: a 250-watt power supply, multiple fans to manage cooling, and a mélange of 12-inch-by-12-inch circuit boards that served as the nerve center of it all, manipulating and controlling a constant flow of data streams. At the backside was an "F connector"—the threaded male protrusion found on the back of a TV set or a VCR—along with a separate Ethernet port. Until recently, it had been rare to come across a single device that featured both connection types. There had been little reason to intermix the two. Now there was. But at the moment that Bud Walters called, neither connection was doing anything. Damage from the flood had knocked out the arsenal's data network, leaving a key US military facility momentarily compromised.

Dick McGarry had an intuition that it wasn't because of any failure of our modem. Rather, exposed cables—the type that had been meant to be used in aerial implementations, not underground—had left the physical infrastructure vulnerable. Instead of being tucked inside sheaths of conduit, some of the cables remained unprotected inside the island's storm drain system. Over time, salt and humidity had corroded the shielding. Now, the storm had overwhelmed the drains, with water cascading into the compromised cables, impairing their ability to carry signals. And because of that, I was apparently going to jail.

On the phone, I expressed empathy and tried to project a sense of calm. I told Walters it was fine if he arrested me, and I'd proceed without protest to the nearest military prison, but I couldn't do so within the time span he'd offered. Even if I caught the next flight from Boston's Logan Airport and hustled down by rental car, it would take me and my tech support team at least eight hours, not the six he'd demanded, to be charged, booked, and outfitted with my jail uniform.

He seemed relieved, I suppose, at the realization I was taking his problem seriously. On the spot, we made plans to hustle down there—myself and a crew of five accomplices representing a third of our entire company. We were rallying the troops because we couldn't afford to lose Rock Island Arsenal. One direct flight and one car rental later, we got to work.

We knew the essential problem revolved around location. But we had to figure out exactly where within the network the cables had been compromised. How? There was no way to clamber down every drain cover to physically inspect mile after mile of cable, looking for lesions. So we came up with an idea borrowed from our experience in Stow: We'd haul out one of our modems, power it up, and connect it at a determined location. From the central station, we'd send a control packet—a sort of electronic sleuth—to our temporary modem. If the packet arrived, it meant that the stretch of cable was working; if it failed, we knew there was a problem somewhere nearby. It was laborious, but it worked. Slowly, we began to isolate places where the trauma had occurred and where new lengths of cable needed to be spliced. Fixing the Rock Island Arsenal network took nearly three weeks, but we'd done our job. Now, the man who wanted me imprisoned loved me.

Except, what about the next flood? The river could be reliably counted on to flood again. We realized we needed a better way not only to ward off physical disaster but to diagnose any number of potential problems before they became severe—not just for Rock Island Arsenal but for everywhere. Here was an important proof point for what was to come. We recognized that if a high-speed information highway was ever going to work at scale, it would need to be able to diagnose itself—to identify troubles before they compromised the entire system.

We returned to Andover and began work on what would become an important part of our technology suite: a first-of-a-kind diagnostic network management system that would intelligently and continually probe for trouble signs—and apply remedies to fix them—before the next network disaster struck. The key to our preventative maintenance breakthrough was the idea of using our own modem to examine parameters and status. It was already there, able to direct traffic, investigate packet behavior, and sniff around for trouble. Two of my software development colleagues, Ed O'Connell and Dale Meyers, got to work, devising principles we would later inject into a first-of-its-kind intelligent network management system for the next-generation modem we were developing from the ground up. Their charge was to replicate via software the heavy lifting we'd done at Rock Island Arsenal. Not just after a failure, but all the time. It was a tangible benefit of working in the real world.

*   *   *

The physics and mathematics of a new data network were etched neatly in pencil across oversized sheets of graph paper, page after page, always in all-upper-case lettering. The fine-grained penmanship was a window into the world of Andrew Borsa, a talented electrical engineer who meticulously conceived, designed, and implemented the sophisticated transmitter and receiver circuitry needed to send data over the cable television pipe at a tremendous scale.

My colleague Andy was an Applitek veteran with an insistent, determined sensibility. He was an inveterate New Englander and a fierce libertarian who had been working on taming the RF environment of Applitek's hardware before I came aboard. Now, with LANcity, he remained a key contributor, responsible for creating and nurturing to life an essential technology that would power our next-generation cable modem. Day after day, he filled his notebooks with the instruction set for a new era.

His work involved QPSK—short for quadrature phase shift keying. It's the signal modulation scheme we had identified as the most reliable way to convince data traffic to play nice over residential cable systems, particularly within the noisy upstream path. Within a typical cable system, this lane of travel, starting at the home and flowing into the broader network, was a spectral no-man's-land, its frequency range of 5 to 42 MHz mostly unused by cable companies—for good reason. "It was like: how do you find a signal within all that noise?" recalled Chris Bowick, a longtime cable engineering executive who worked as the chief technical officer for the Colorado-based cable company Jones Intercable.

Besides the generally noisy spectrum, signal leakage—the beast we called "ingress"—was a big problem. In homes where people had loosely screwed in F-connectors to VCRs and bedroom TV sets, even a home hair dryer or a vacuum cleaner could compromise an upstream signal by introducing more noise into the network. QPSK looked to be the way out of this mess.

For years now, Andy had been laboring to get our iteration of QPSK working and functioning reliably. It was demanding work. A notebook page labeled "Channel Filtering Requirements," one of hundreds filling up a set of thick binders, addressed one small sliver of Andy's explorations. Here, he handwrote dozens of mathematical calculations describing the behavior of traffic flows and the arithmetic behind them. At the margins and on top of pages appeared neatly penciled reminders of aberrations he was working to tame. "Could conceivably be done with a 5-pole + 2 notches," one notation pointed out. "RX only needs to reject center of adjacent channel by 37dB," instructed another. Page after page, Andy was documenting a seemingly bottomless well of data points and instructions, theorems and commands.

Andy and his engineering compatriot Paul Nikolich had been all-in on QPSK for several years, chipping away at what sometimes seemed to be a never-ending cascade of issues. The work could be punishing. Complex arrangements of corrective filtering remedies that would ward off signal degradation required a delicate courtship of physics and mathematics, occasionally bordering on art. Our printed circuit boards were equally our enablers and our tormentors, expensive and time-consuming to fabricate, test, and align. It took us a full day to properly tune a new modem, undertaking more than 40 separate alignments before declaring the device to be ready. Even after painstaking design work, construction, and testing, the outcome often would be frustrating. Fluctuations in temperature could send the device out of whack once again, requiring on-site doctoring from a technician.

Paul knew the terrain as well as anyone. The son of Yugoslavian immigrants, he grew up in Queens, New York, where his father, a Con Edison worker by day, ran a radio and TV repair business during off hours. Surrounded by the scattered detritus of a booming electronic age—vacuum tubes, speakers, and other components—Paul displayed an engineer's native curiosity from an early age, tinkering with bicycles and model airplanes and helping his dad bring dying electronic devices back to life. It was hardly a surprise, then, that after graduating with engineering and biomedicine degrees from Brooklyn's Polytechnic University, the baritone-voiced New Yorker found his way into the orbit of electrical engineering, settling in with Applitek to work on digital communications systems.

As a basketball power forward at Polytechnic University, Paul had set a record for rebounds in a single game at twenty-seven, a mark that held decades later. Having gotten to know him, I could see why. For LANcity, he brought a singular sense of purpose and determination to his work in this new, unexplored arena, writing custom software from scratch to attempt to simulate the conditions we would encounter in the real world. "All hand-written code," he later recalled. "We were doing software even though we weren't software engineers."

By early 1990, in alignment with our Contra plan, we were beginning to work on taking the same QPSK principles we'd applied in the early era and translating them to a second-generation modem technology that would do away with much of the laborious tweaking and aligning of analog circuit boards. The transition from an analog QPSK modem to a hybrid analog/digital architecture would be vital. Without it, the deployment of low-cost consumer modems that could withstand the vagaries of the residential cable television environment would never have occurred.

Paul Nikolich, Andy Borsa, and later, my former GE colleague Bill Corley, would go on to work for nearly five years on the intricacies of our modulation schemes and surrounding technology. Five years of brain-crunching, meticulous development effort from a team of extraordinary RF engineers possessing a deep understanding of physics, math, software, and electromagnetic principles. Our collective brainpower was a big reason for my confidence that we could get our mass-market, residential cable modem built and perfected before anyone else.

Andy Borsa would later serve a single term in the New Hampshire House of Representatives before succumbing to health challenges triggered by a skiing accident. His notebooks remain tucked away in my private library. Inside them are the inscriptions of a game-changing technology, one that was about to captivate one of the most powerful players in the booming computer industry.

CHAPTER 5

THE MISSING LINK

We knew that if any one of us stopped paddling, the whole boat would sink.

The lessons from our Stow experience reinforced what I already knew. We were running fast toward the limits of our first-generation modem's capabilities. The bucket list was calling: We had to develop the next generation of technology.

To do that, we needed an infusion of money. To be sure, we were proud of our "earn to grow" mentality. We'd revitalized a moribund operation, elevating monthly revenue to $200,000 by focusing on service and support fees and occasional sales of cable modem starter kits. We were making enough to pay our bills. But at this rate, we were never going to be able to fund large-scale development of advanced technology.

What to do? The conventional path was to call on venture capital investors. But the investors backing our company had stopped funding our operations nearly four years earlier. As for fresh investment possibilities: Nobody wanted to adopt a nine-year-old puppy. Besides that, few people within the venture capital trade or the cable industry itself subscribed to our core vision: that we could spark a transformation, elevating cable television from a video delivery orientation to a powerful agent for the information revolution. The few venture capital investors I spoke with offered up mostly polite stares. Discouraging signs were common. At a 1989 academic conference where I spoke, one attending Ph.D. likened cable television to a sewer system in terms of its ability to manage data traffic.

The general lack of faith left one avenue: a strategic partner. And there was little question about who that might be. All along, DEC's Jim Albrycht and John Kaufmann remained believers in our shared vision. Our first-generation modem was still sitting idly in Kaufmann's living room, but the two friends kept the faith. The signal trace etched into being at a Boston hotel was still alive.

DEC features large in the history of computing. The company's CEO, a Massachusetts Institute of Technology graduate named Ken Olsen, was an engineer at heart, captivated by the new possibilities of the computing age. In the rearview mirror of the US high-tech industry, Olsen was often criticized for missing out on a personal computer revolution that would come to be dominated by the likes of IBM, Apple, and Commodore. DEC, wedded to its aging VAX "super-mini" computers, was left mostly standing on the sidelines as the home computer category roared. In 1983, IBM sold more than 750,000 personal computers. DEC sold 69,000. Olsen had remained doubtful about the trajectory of personal computers at large scale. Rather, he believed, the world would continue to depend on the sorts of time-sharing computer systems that were core to DEC's business. But the script played out differently: Computers became ever more powerful, and fast broadband networks spurred a reinvention of Olsen's concept of centralized computing in the form of today's cloud computing environments. In the end, DEC itself was acquired by a personal computer maker, Compaq Computer Corp.

But calculations about the future of computer hardware weren't the only subject consuming DEC in the early 1990s. Happily for us, DEC's interest in metropolitan networking was also growing. Through a business unit called Advanced Networking Technologies, the computer giant was exploring ahead-of-their-time technologies like wireless metro-area networking and fiber distribution networks. I loved DEC's technology-first orientation, a culture steeped in a spirit of innovation. John Cyr, a DEC engineering manager who would come to play a major role in LANcity's advancement, characterized the DEC philosophy in the early 1990s this way: "We really like cool stuff, and we really don't care if it is going to make much money." At least, he qualified, not immediately.

What we were attempting to do at LANcity surely qualified as "cool stuff." But there were serious reservations within the DEC leadership about whether we could make it happen. We learned that the hard way in July 1991 during a technical presentation at DEC's Littleton, Massachusetts, offices.

William Hawe, "Bill" to friends, was an Ethernet icon, his name spoken reverently alongside those of Robert Metcalfe, the legendary Ethernet cofounder, and Metcalfe's partner, Dave Boggs. In the early 1990s, Hawe was the chief technology officer for DEC's networking business; he would later be named the CTO of the data networking company Bay Networks and its successor, the Canada-based tech giant Nortel. In other words, he was a major player in the Ethernet kingdom. If Bill Hawe was in the room, important things were being talked about and important decisions were being formulated. On this day, Hawe was indeed in the room. And what he said was a punch to the gut.

My colleague John Ulm and I were guests at a theater-style conference room at the company's Littleton, Massachusetts, building. I felt nervous and overmatched from the start: Dozens of high-caliber DEC engineers on one side, John and I on another. For nearly four hours, starting at 2 p.m., John and I patiently reviewed our technical roadmap, explaining our work across multiple fronts to create a technology recipe that could unleash the power of the data-over-residential cable television network—and forklift high-speed data networking from the office to every home in the metroplex.

The mood was tense. Engineers immersed in the intricacies of data networking seemed skeptical. They hardly spoke. "Very, very intimidating," John recalled. "Here I am, a relatively young engineer, with this guy who was a king of Ethernet. We present all our ideas, and we get to the end of it."

And then, the verdict: "Bill Hawe gets up and says: 'This is too complicated. It will never work.' And he walked out of the room." One of the captains of data networking was declaring our work to be impossible.

* * *

John Cyr was a sharp-minded engineering manager who worked in DEC's Networks and Communications division. Until DEC had begun exploring a possible deal with LANcity, most of his familiarity with cable television had been as a viewer, enjoying live sports on ESPN and the regional sports network NESN. Now, Cyr was turning attention to a different aspect of the cable equation. DEC had tasked Cyr with determining if LANcity's vision and product set, and more broadly, the entire cable-broadband dream, was real. Cyr hit the road to take it all in, traveling to conferences and conventions in the cable industry's annual calendar.

In a 1992 memo to superiors, he encapsulated the uncertainty of the moment. On one hand, he was excited by the cable industry's growing interest in new technology and networking architectures. But on another, he expressed worry about the payoff: Was there really sufficient demand in the United States for the type of high-speed broadband network we were working to create? Lately, cable companies and their competitors in the telephone business, Cyr commented, had been "putting a lot of faith in the 'Field of Dreams' concept: let's build the infrastructure and the applications will come." (The iconic movie Cyr referenced had come out three years before.)

It was a valid point; when Cyr published his memo, the Internet was unfamiliar to most Americans. The so-called network of networks was used primarily by universities, researchers, and government agencies, along with a few early hobbyists. The first website had been published only months before, in August 1991 (by CERN, the European Organization for Nuclear Research). Despite the inroads made by IBM and Apple, only about 22 percent of US homes had computers. The world was a long way from the forthcoming "dot .com boom."

Beyond these marketplace boundaries were engineering concerns. DEC, a demanding company from an engineering and technology standpoint, would have to swallow some hard realities. With cable television, there were imperfections and workarounds requiring compromises involving signal delays, noisy upstream paths, and overall network reliability. Among other things, Cyr noted that the cable industry tolerated a higher degree of signal erosion—measured by way of a metric known as a "bit error rate"—than was typical in the world of digital LANs.

Finally, there was another possible obstacle lurking: Leaders in the US Congress were considering new regulatory policy surrounding the cable industry, and worries were growing among industry executives and lobbyists that the outcome could be negative for the industry. By June 1992, a bill promising to reintroduce government oversight of cable rates had made it to the floor of the US House of Representatives, signaling that lawmakers were serious about re-regulating cable TV.

In the end, though, Cyr concluded the opportunities for DEC outweighed the concerns. "(DEC) needs recognition in this industry," he wrote. "We need to help in this revolution." And here we came: a small, nearby company with an idea that could conceivably rocket DEC faster than anybody else into the world of city-wide data networking. Even if Bill Hawe said it couldn't be done.

Cyr recommended that DEC deepen its involvement by lending engineering expertise, offering counsel, and, subject to meeting a rigorous schedule of milestones, supporting us financially. By June 1992, we had to have our detailed design specification completed. Around three months later, in early October, we needed to complete a large-scale simulation of the integrated circuit that would power the protocol for our forthcoming second-generation cable modem. By March 1993, we had to have a fully operational prototype ready, and by August 18, we needed to ship real cable modems to real customers. The breakneck schedule was bound by hard-wrought covenants. Miss any of the major deadlines and DEC had the right to yank its funding and discontinue our partnership.

Getting a deal done would be a painstaking process. From the day I met Albrycht and Kaufmann through early 1992, I participated in a head-jarring cascade of 239 meetings with DEC staffers. Some weeks, we'd meet daily, sometimes twice a day. Every time it seemed we were capturing interest from one department, we'd be introduced to another. Patiently, I'd retell the story of how our cable modem could be the linchpin of a new networking era—at a massive scale, connecting millions of residential users. New acquaintances from new DEC departments would nod, ask the same questions I'd been asked dozens of times, express enthusiasm, request engineering documentation, set up follow-on meetings, and then suggest we talk to so-and-so in another of DEC's seemingly bottomless arrangement of divisions, teams, and subsets.

Then finally, daylight: A memo from DEC's legal team spelled out an alliance between LANcity and DEC that would inject $2.5 million into our company over a period of 18 months. We had been clinging to life, waiting for a breakthrough that would support the dream. Now it was within view. A memo circulated within the offices at DEC's headquarters declared that we could bring a nearly three-year head start over any competitor in the data-over-cable arena.

Even so, corporate life moved at its own sluggish pace. Only after I sent an urgent letter warning of imminent financial peril was the contract finally signed in February 1992. The terms called for seven payments, with the first amounting to $626,000 and the rest based on our ability to meet technical development hurdles and deadlines.

Now, we could get down to business. Two especially enthusiastic members of the DEC team were Kaufmann and Albrycht. We met often at DEC's Littleton office or around a cafeteria table downstairs at our building on Brickstone Square, where, in the early 1900s, a thriving mill district had been situated near the Shawsheen River. I made it a point to keep a supply of gummy bear candies handy when we got together. Either that or I'd invite the two men to one of our evening dinners prepared by my mom.

Always, the two listened intently. They pressed me about how cable television could overcome the limitations of the twisted-pair telephone network. They demanded to know about detailed field engineering data and specifications. Every time we made an engineering advance or tweaked our technology, they were the first to get the story. Over time, we developed a close brotherhood, a fellowship driven by our mutual interest in what LANcity was up to. I knew we were getting somewhere when we began completing each other's sentences.

It was our persistence in Stow that ended up spurring a deal. Nick Signore and a technician had been painstakingly upgrading and adjusting the return amplifier modules at the Nashoba Valley Cable system. It was arduous, time-consuming work. Some of the bolts holding the amplifier casings together were badly corroded, resisting attempts to open them. Nick and his Nashoba Valley colleague kept at it, amplifier after amplifier, adjusting a familiar quartet of signal attributes—gain, distortion, noise, and frequency. We were grooming this short span of the Nashoba Valley Cable system with the delicacy and care of a doting parent.

With Nick assuring our team that the amplifier cascade was ready, we gathered again in Kaufmann's living room. Nick stood to my side, Jim Albrycht hovering excitedly nearby. It was showtime. And this time, a different outcome: We watched as a computer file that had originated at Albrycht's home was instantly downloaded to our connected workstation.

It was our eureka moment. We had demonstrated that with the right alignment and adjustments around the residential cable television infrastructure, it was possible to deliver data at high speed over a residential cable system, where traffic would flow over the same wires that carried the signals of ESPN and CNN. If we had champagne handy, we would have doused ourselves with the stuff. We made do with handshakes and high-fives.

It was all the more satisfying because we had done it the LANcity way: However we could. We worked overtime. We borrowed advanced workstations from DEC. A friend at Hewlett-Packard had loaned us a spectrum analyzer we could never afford. I'd rented the cheapest utility vehicle I could find. We convinced Nashoba Valley Cable to lend us an experienced line technician and offer an unused signal path to characterize and model.

There were additional proof points. My DEC friends were excited by the results of a project involving an array of information kiosks that had been set up at Boston's World Trade Center for DECWorld, a large developers' conference. LANcity had connected the kiosks to a dedicated institutional network engineered by Boston's cable company, Cablevision Systems. Later, we expanded our original Nashoba Valley Cable implementation, rolling in a few more homes where DEC employees lived and connecting DEC's nearby offices in Littleton, Massachusetts. The second time was the charm. In a written report, Albrycht noted that this time, we'd gotten the connection running in a matter of days.

There was also a more personal influence that carried weight at the corporate level. DEC's CEO Olsen had been dealing around this time with his wife's serious illness. I learned from my DEC colleagues that Olsen had been carrying computer storage disks containing radio-imaging files from one Boston hospital building to another. In doing so, Olsen was experiencing first-hand the shortcomings of our modern computer networking systems. We could get large files transmitted across an interior office environment using Ethernet, but outside, in the broader world, our network systems were wanting. Although technologies like FDDI (for fiber-distributed data interface) were in place in a few locations, the fiber-optic cabling was expensive and the deployments sparse. As Olsen was discovering, the only sure remedy was "sneakernet," a not-entirely-flattering term for physically carrying a disk or storage tape from one place to another. I never met Olsen in person, but Albrycht had kept the CEO and his executive team up to date on our mutual explorations through a detailed quarterly report titled *Community Networking Update*, which he wrote and hand-delivered to DEC executives. He assured me the DEC leader was intrigued.

Before signing the partnership deal with DEC, I wanted to make sure my team was on board. The only major move I'd made on my own was to use my home as collateral for our early investors; I didn't want anybody else risking theirs. John Cyr would later remark that I ran the company more as a father figure than an employer. But I could be a demanding parent. I wanted to hire people who would devote the next three to five years of their lives to our vision of a connected world. I also wanted people who wanted to be challenged. The vote from my team was emphatic. Not "yes," but "hell, yes." The team saw the merits of the DEC alliance immediately. We would establish a partnership with a major computing power, a company that had a passion matching ours for, as the late Apple cofounder Steve Jobs would articulate, putting a "ding in the universe." It was the first glimmer that we weren't just some small, fringe tech company dabbling on the outskirts.

With that first payment from DEC in the bank, LANcity had an injection of new energy. We could turn our attention fully to realizing the broadband dream. I would need to buy more gummy bears, this time in extra-large packages.

* * *

By now, DEC's engineers had become well-acquainted with our first-generation modem, modified from Applitek's original product. They called the thing our "boat anchor," alluding to its nearly 80-pound weight. DEC insiders, accustomed to large-scale manufacturing for its own computing systems, marveled at the homespun efficiency involved in producing the NI-10E. No huge assembly lines, robotics, or to-the-second manufacturing routines for us. Instead, a team of four individuals at LANcity was responsible for assembling our components, making final alignments to delicate circuits, testing the finished product, and readying it for shipping. I frequently volunteered for the shipping part of the job, yanking down the metal door of the nearby elevator and latching it securely to begin the slow ride down four floors. The process was worlds apart from what DEC had known.

Our business relationship went by the acronym OEM, for original equipment manufacturer. Under our arrangement, LANcity would continue to design and manufacture our proprietary cable modem, but now DEC would marshal its marketing, sales, and services resources to give the technology wider exposure. DEC had been calling the effort "Ethernet over TV," or ETV. Later, DEC would come up with the catchy name of "ChannelWorks."

For our forthcoming modem, built for commercial users and named LCB (the "B" alluded to business customers), the goals were ambitious: reduce the weight to 12 pounds from the current 80 or so while expanding the distance over which we could send and receive signals to at least 160 miles (and later, 200). In addition, we were determined to slash the list price to $5,000 from the previous $18,000. To give our cable industry partners technical leeway, we needed to expand the range of frequencies we supported. Our production volume would have to expand, too. We were planning to produce up to 30 second-generation cable modems per day, compared to one per day for its predecessor. At the same time, we would reduce the requisite number of alignments from 40 to 2. Backing this wholesale revamp would be a switchover to digital signal processing. Then, the secret sauce and crown jewel: our Unilink-II media access control protocol.

"It was a paradigm that nobody had addressed yet," recalled John Ulm. John had been working nearby, in Chelmsford, Massachusetts, for Apollo Computer, a developer of networked workstations with impressive graphical capabilities. Apollo's powerful machines held their own against those of category giants like Sun Microsystems and Hewlett-Packard. Scientists and engineers used them to support big research and product development ambitions. These were projects that—this was important in our world—involved moving huge amounts of digital data. John had developed an impressive facility for wrestling with round-trip signal delays that are the familiar bane of data networks. If we were going to rein in these gremlins at a massive scale, both distance-wise and user-wise, I needed someone with his talents and knowledge. So I made the pitch, but it was not to be. After being introduced to me by a mutual acquaintance, John politely accepted my phone call and, just as politely, turned me down after a brief conversation. To him, I was a nobody: just one more startup company founder who spoke with an accent. But then, I got lucky.

The day after we spoke, John learned his company had agreed to be acquired by Hewlett-Packard Co. He and his team would have to move to Sacramento, California. For John and his wife, both hard-wired New Englanders, a cross-country relocation was a nonstarter. Suddenly, LANcity was back in the picture—with one hurdle remaining: To convince him to take a job with LANcity, I had to convince Cheryl, his wife. She visited me in person at our office in Andover, reminding me that John had already dealt with one start-up employer's demise and laying down the terms: no hundred-hour workweeks. Don't go bankrupt. Find a way to survive.

What was I going to do? Negotiate? I needed John Ulm more than John Ulm needed me. I agreed to Cheryl's conditions and added a bonus: A bottomless supply of M&M candies in my desk drawer for their three young boys, who would become welcome fixtures at our offices as we grappled with the complexities of building a new, more powerful cable modem powered by our new protocol.

We called the successor protocol Unilink-II, but it was similar to the original in name only. We started fresh in order to overcome vexing issues around the scale of deployment, quality of service, compatibility with real-world cable systems, and improved efficiency on longer cable plants. It had to optimize the transport of Ethernet packets that by now had become the dominant LAN protocol. John Ulm appreciated that this protocol had proven its chops in short-span environments. But in a metro topology, he knew the technology ran into serious issues. He likened Ethernet's distance challenge to accommodating a conference call with people across the globe, where the combination of signal delays and lots of interruptive blurting by many participants makes it hard to sustain a sensible conversation. "Now add hundreds of additional people to that call and then stretch it to where some people are on the moon," he cautioned. "It becomes utter chaos."

Taming the chaos would become the main objective for John and LANcity colleagues Ed O'Connell, Cindy Mazza, Gerry White, and Chris Grobicki. Their names adorned a 1993 US patent describing a protocol for a "Communications Highway Network System" that attacked concerns around how network resources would be allocated and how packet timing would be accommodated. One of their foundational decisions involved partitioning the bandwidth coursing through cable systems into slots, using techniques of time division multiplexing (or TDM)—a way to allow multiple users to take turns using the same communications channel. Keeping these assignments orderly was the job of a pacer that lived within the network, synchronizing an interplay of users and data flows.

Then, an injection of engineering genius. For each available TDM slot, the Unilink-II protocol allowed for a dynamic, either-or mix of data traffic treatments. Depending on network conditions, usage contention, and overall demand levels, different users and different applications would be handled in different ways. Some users and some applications would be granted dedicated bandwidth reservations offering contention-free throughput, even under heavy traffic conditions. Others would latch onto whatever amount of bandwidth happened to be at the ready, providing immediate access and low latency. Bandwidth-intensive, mission-critical applications like phone calls or media streaming would enjoy the former treatment. Email and other more "bursty" applications could get by with the latter. This dynamic pairing of possibilities was the heart of Unilink-II.

One industry engineer who would come to admire the concept was Tom Moore, a pioneer in the satellite-delivered Internet sector who worked on early trials involving cable-broadband technology and the global standards that flowed from them. "The thing that LANcity did so beautifully was to combine those two concepts," Moore said in a 2015 interview conducted for CableLabs, the industry research and development organization. "It was an elegant solution."

Parallel to our Unilink-II development work was a determined embrace of existing technology. We had already adopted two essential standards: the IEEE 802.3 Ethernet packet standard and the cable industry's 6 MHz channel allocation scheme. Now, we were integrating newer software standards as well, ideas that ran counter to what some other big tech companies would end up theorizing. We became convinced the core language of the Internet—TCP/IP—could hitch a ride over the networking industry's Ethernet data link layer. The basic premise of the TCP/IP protocol is that information is busted into packets—the elemental bits of Internet traffic, tiny slivers of messages that course through data networks—and then transmitted and reassembled on the other side.

We integrated a simple network management protocol (SNMP) in order to allow our customers to choose from multiple plug-and-play vendors whose systems ensured a smooth-running network and easy diagnosis of problems. Standards were the right path; one we were determined to pursue if we wanted any chance of getting to our desired price point for the new modem. But shortly after LANcity came into being, we needed something equally important: the right people.

* * *

With funding set aside for second-generation design and development, I began building out my team. I convinced Bill Corley, my mentor from the GE days and a brilliant engineer, to join LANcity as a full-time employee. Paul Nikolich, the electrical engineer who had joined Applitek in 1986, had taken on a new job but agreed to continue as a key advisor, helping us overcome tricky RF issues and exchanging leisurely New England Saturdays and Thursday evenings for pizza sessions where we'd whiteboard engineering problems and explore ways to address them. I recruited a fast-thinking technical advisor and mathematician, Dale Meyers, who had developed the original operating system for the earlier Applitek product line.

It was a bit like casting for a movie: The more talent you bring aboard, the more people want to play a role. Chris Grobicki, the ex-Applitek colleague and design engineer who left shortly after my arrival, agreed to contribute as an advisor (and later an employee) out of intellectual curiosity and growing interest in the vision of data over residential cable TV. As a developer of the original Unilink specification, Chris remained convinced important breakthroughs could be realized in the marriage of high-speed data and cable television systems.

Chris was a smart, studious engineering manager with a soft-spoken demeanor and a keen appreciation of scale—able to make technology work reliably at extremely high levels of user demand. Born in Montreal, he'd been transplanted as an eight-year-old to Knoxville, Tennessee, in the early 1960s, whipsawed by the move to a then-segregated southern United States. Like Nikolich, he'd been smitten at an early age by the workings of electronic devices, tinkering with amplifier tubes and power supplies, poring over technical manuals, and becoming determined to follow in his dad's footsteps as an electrical engineer. As a student at the University of Tennessee, he worked as a lineman for a regional utility company, Florida Power & Light, where he earned an industrial badge of honor by strapping on cleats and climbing poles. I'm convinced that a secret power of LANcity was our up-close-and-personal familiarity with the reality of the cable infrastructure, creosote-stained utility poles and all. Chris, importantly, also possessed a down-to-the-bones knowledge of the original Unilink protocol. As employee no. 4 at Applitek, he'd focused on implementing the firmware and hardware tied to the technology.

Also sticking around was my colleague Andy Borsa, who had been working the demanding terrain of QPSK modulation for nearly five years. Andy stayed on, insisting on his own terms and schedule, as a consultant and invaluable collaborator: a studious, insistent, smart mathematician and engineer and very possibly the world's leading QPSK developer. Shortly, I would bring aboard an iconoclastic coin collector who would catapult our company into an entirely new stratosphere.

These newcomers worked side-by-side with my 13-person LANcity all-star team to make up our pizza crew: a mix of new employees, advisors, hired contractors, and just plain interested technologists. There were 20 of us in all. We'd pick up pizza from a local parlor and serve it upstairs in our conference room. With slices in hand, we worked to identify the biggest, most daunting engineering tasks we could. We listed them on oversized sheets of paper taped to a wall as we attempted to get a grip on the endless list of technical challenges we'd need to resolve. By the time there was no more room on the wall for another sheet, we knew the evening's work was completed, and we should go home.

I tried to set a tone of optimism even when the obstacles looked insurmountable. A few months in, Paul Nikolich came to me with a sobering confession: "Rouzbeh, I hate to tell you this," said the man who would become my closest friend. "But this cannot be done. The cable industry is never going to be serious about data. You're killing yourself for nothing." I let him know I understood his frustrations but convinced him to keep pressing forward, which he did.

For LANcity, the focus was on high-quality technology. We weren't good at brochures or product literature. The only person on our team who could write passable copy or manipulate desktop publishing software was Les Borden, an energetic right-hand man. He doubled and tripled as a technical writer, customer-service agent, and traveling companion at trade shows. It was Les who accompanied me on a memorable trip to Nashville at a bottoming-out point when we were days away from missing payroll. The cheap motel where we'd reserved a room was like a scene from a pulp novel: Getting a room key meant pushing bills and coins underneath a bullet-proof glass shield at the entry counter. That night, gunshots from outside our room sent Les and me scurrying. We slept in the car at the trade show parking lot.

Les was essential, occasionally working side-by-side with me for periods of 24 hours straight, sometimes longer. He was adept at LANcity diplomacy, able to coax from overtaxed engineers technical documentation details he'd squeeze into customer manuals I'd tasked him with writing. I'd supplied him with an aging PC-emulating workstation loaded with an Intel 80286 processor and WordPerfect software. It was all we had.

Around the time our DEC partnership materialized early in 1992, our technology was starting to be noticed. Occasionally, we'd answer phone calls from people in the cable industry who had heard of us. Were we the company that had an Ethernet bridge for cable systems? They'd usually ask if we could put them in touch with a distributor—an intermediary company that stocked and resold our technology. We had no such thing; anything we sold, we sold direct. But I'd assure them my lone sales representative, Tony DiSessa, a gregarious conversationalist from New Hampshire, could fly out for a visit within the next day or two. Sometimes, I'd make the journey myself.

I'd given everything I had to bring us this far. At 34, I was unmarried, had no children, and barely slept. I was willing to do anything to keep our company and our dream alive. I contributed to whatever the day's agenda presented: engineering, manufacturing, sales issues. I hauled heavy modems down the utility elevator that led to our shipping dock. I mediated personal disputes, including a tense spat between one of our engineers and the building janitor; after one especially long workday, they'd become embroiled in an argument. In this instance, I sided with the janitor. Like everyone, he deserved to be treated respectfully. Jerry Amante, a colleague from DEC and a longtime IT engineer, would later call me "the hardest-working individual I've met during my career." Coming from Jerry, a demanding realist who'd grown up in the hard-scrabble environs of nearby Lawrence, Massachusetts, that meant something.

My reverence for the hometown Boston Celtics had proved to me that you need a deep bench to win a championship. I had one. LANcity veterans including Ed O'Connell and Cindy Mazza, intuitive and fast-thinking software engineers, had remained on the team, energized by the chance to work on breakthrough technology. Dale Hokenson, a colleague who had worked on a Cray supercomputer at Colorado State University, joined us to focus on customer relationships. In a backroom was our long-haired, soft-spoken quality manager Mike Sperry, a Proteon refugee who oversaw a grab-bag of duties, including applying for regulatory approvals of our technology. In a look-back video I would later produce, Mike credited LANcity's success to "a little bit of technology, and a leader who was crazy enough to believe he could actually turn it into a company."

Working closely with Mike was an army of two: Sal Privitera, a Rhode Islander who arranged and oversaw our technical and product/design testing operations, and Craig Brinker, our materials purchasing manager who lived on Cape Cod but rented a nearby hotel room during the workweek. Gerry White, a software architect who'd begun his career at Wang Laboratories and, at John Ulm's recommendation, I'd recruited as our chief technology officer, brought an irrepressible English accent and an unmatched facility for solving software puzzles. In the space of only weeks, Gerry built out vital parts of our infrastructure: He'd devised a network management system that would probe the workings of our technology to sniff out problems before they became trouble. He assembled a first-class software toolset for our development platform.

I promised my associates better days and more satisfying work were ahead as we plowed new ground with work on our next-generation modem. In the opening paragraph of a memo I circulated in March 1992, shortly after we deposited our first check from DEC, I spelled out the grand vision. "During the next 17 months," I wrote, "we will be taking over a major responsibility that can revolutionize the way we live, work, and even raise our children." I outlined the marketplace potential, suggesting we could usher in the fastest billion-dollar market on record ever for the telecommunications industry.

Even with our talent additions, however, we remained a tiny crew relative to the brash objectives we were pursuing. I invested every new dollar back into the business and, even then, recognized we'd need to double, or triple, the expected productivity any one employee could realistically be tasked with. The same held true with financial efficiency: We did whatever we could to get $3 worth of return from every $1 we spent.

The key to achieving that was devotion. We often spoke about multiplying the contributions of one person into the equivalent output of three. As we ambled into the office in the morning, firing up computers, prepping signal display monitors, threading spools of coaxial cable from one computer to another, my team seemed to display a sixth sense about teamwork. We knew that if any one of us stopped paddling, the whole boat would sink. Knowing this, and knowing there is a tight alignment between work and life, I would listen carefully to whatever my colleagues had to tell me. I heard about marriage difficulties, financial strains, spiritual questions, and worries about kids. I counseled one member of our team, a sensitive soul, in repeated cubicle confessional sessions, taking time whenever called on, regardless of whatever crisis was unfolding. It didn't matter what the subject was; I was there to be an ally, to rally the troops, to keep the focus tight, the spirits high, and the emphasis on going forward. My colleagues made for an opinionated, noisy, rollicking bunch who believed passionately in faith, in humanity, in religion (or its absence), in our technology—and in the New England Patriots. They weren't afraid to try to convince anyone of anything. The passion transcended mere lunchroom chatter—if you could call this heated conversation chatter. On some days it sounded more like a revival meeting or a raucous political debate. There was plenty of gamesmanship, too. Les Borden often would goad a fellow karate black belt, Ed O'Connell, into lunchroom sparring matches. Les doubled as one of our resident experts in darts, daring anybody to try to beat him in the lunchroom. Nick Signore, our man from Stow, was the only person who did. (Repeatedly.)

At least once a week, I'd greet my parents downstairs and help cart up platters overflowing with elaborate, delicious Persian dishes. The well-fed team is a happy team, that much I knew. And being happy was important. We were working like mad. "It was hard work, around-the-clock work," remembered Nick Signore. "But we were doing interesting stuff."

My own work habits were all-consuming. My habit was to drive home for a few hours of sleep after midnight and return early in the morning. One morning, around 1:30 a.m., driving just north of Merrimack College, I was broadsided by a car that spun me around like an amusement park ride. I collided with a utility pole, barely aware of what had just happened. Another driver, a young woman on her way home from the campus, had slammed sidelong into my Honda Civic, sending me flying. Somehow, I'd shot across the passenger seat, escaping serious injury from the collision that totaled the car and might have squashed me otherwise. To the responding North Andover police officer, I looked the part of a malfeasant: bleary-eyed, unshaven, harried, and confused. He assumed I had been drinking alcohol. The young student who had caused the accident took pains to convince him I wasn't at fault. I spent a few hours at the hospital, caught a ride home, showered, and returned to work later that morning.

My challenge to the group was to make good on the original ambitions I had committed to paper back in 1988 at my colleague Mary Lautman's house. Slash the price. Extend the distance. Support simultaneous users at an enormous scale. Make sure we can adjust on the fly to a broad range of operating conditions. Also, shrink the size. Harvest the QPSK work we had been grinding away on for years. Morph from analog to digital signal processing. Complete the specifications for our Unilink-II protocol. Make easy installation possible, transforming the notion of "plug and pray" into "plug and play."

And one more thing: replace the elaborate physical interplay of multiple, oversized electronic circuit boards and their analog components with a new type of processing intelligence. We passed around pizza boxes on a Thursday evening. "Let's build a chip," I said.

CHAPTER 6

WHEN DOVES FLY

*We weren't trying to perform dark magic.
Math and physics were on our side.*

The assessment from DEC's Bill Hawe about the challenges involved in implementing our MAC protocol in an integrated circuit felt deflating, to be sure. But we refused to surrender. We were too far along, intent on getting to the finish line. We were also certain we had a valuable head start, years ahead of any competitor.

Part of our conviction was rooted in our knowledge of mathematics and physics. Part of it drew from our history of field deployment experience. And part of it depended on faith. There would be moments when I experienced something almost spiritual, a guiding hand tied to an essential calling.

John Ulm felt it, too. "My entire career, doing Token Rings, and then the startup, it had all come together," he said. "It was like I was fated to be here, at this moment, to do this design. Everything I had done previous to this led to this moment." To that point, the expatriate of Apollo Computer and father of three M&M-gobbling boys considered Bill Hawe's assessment not as a death knell but as motivation. "That threw down the gauntlet," he said. "That was my motivating factor. I'm like: 'No. We're going to make this happen.'"

Privately, I had to admit there was at least some basis for Hawe's skepticism. Inventing a new protocol and translating it into the form factor of a new integrated circuit from the ground up—tossing aside the work and the systems that powered Applitek's first-generation technology—meant dealing with hundreds of problematic technology and practical deployment issues. Developing a wholly reinvented successor to our Unilink protocol alone would be all-consuming. Nobody had invented a media access control platform that would work at such immense scale, accommodating big spikes of concurrent demand over a network invented to distribute TV channels, not waterfalls of data.

But that was just one challenge. Beyond our MAC protocol work, building in frequency agility was among the more pressing demands. To align with cable industry practices, we needed to allow cable companies to assign our data path to whatever 6 MHz channel they happened to have available. We knew that intuitive, self-installing software with auto-upgradability had to work flawlessly. In addition, spectral purity was a big concern. Within our 6 MHz lanes, we had to make sure our data signals would not leak into nearby video channels or overpower the inputs to the amplifiers in the path to the headend.

Besides that, almost everywhere we looked we saw gremlins known as "impedance mismatches." The term referred to a predicament that happens when an F-connector that's loose, or left dangling, causes an echo in the network. When that happens, the modem becomes confused, not knowing which signal to process. Across the vast infrastructure of cable television in the United States where millions of F-connectors had been screwed into millions of TV sets and millions of videocassette recorders, micro-reflections caused by impedance mismatches threatened to be prolific. We had to add to our laundry list some mechanism for compensating for them in mid-step, without anybody ever noticing.

There were other operational challenges. Cable industry technicians would have to be able to install our modems without incident in home offices, bedrooms, dens, and kitchens. Lately, the cable industry, under siege from federal lawmakers, had been doubling down on efforts to overcome a sometimes-sketchy record for customer care. Now, with competition rising and regulation looming, the standards for customer care were elevating alongside them. We knew we'd have to play along nicely, supporting everything from positive customer interactions in the home to integration with back-office billing systems that had long been a headache for cable companies. We'd also have to support a commitment to tender loving care for people's prized computers. The last thing our cable partners would tolerate was ruining somebody's new PC.

Sorting through these considerations meant relying on rigged-up networks of our own. Inside LANcity was a replica of a mile-long cable system. We had connected a series of amplifiers over a long cable run to dozens of our cable modems, using a data traffic generator to mimic the flow of packets across a residential cable. In a backroom sat oversized wooden spools wrapped with cable we'd purchased on the cheap from cable industry friends. Similarly, we acquired amplifiers wherever and however we could. New, off-the-shelf amplifiers—or "cans," as our cable friends sometimes called them—could cost more than $5,000 apiece. Thanks to some connections in the industry, we were getting them for fractions of that. We purposely mixed newer amplifiers with older iterations, knowing we'd encounter both varieties in the real world of cable television.

Our lab was an early version of similar setups that would later be used to help devise a global standard for cable modems. But it wasn't an exact replica of reality. We knew we weren't yet facing what would surely be vexing environmental pressures: severe temperature fluctuations, birds nesting on wires, amplifier settings going rogue. That's what our work in Rock Island Arsenal, in Stow, and in our broader list of real-world deployments was for. But for the moment, this is what we had: spooled reels of coaxial cable connecting cable modems to data generators, suitable for determining whether packets could make the journey with acceptable levels of performance.

The challenges were unending. But there was hope. I had not heard of a single issue that defied the laws of physics, or mathematics. Every problem, as troubling as it may have sounded from a development point of view, was solvable. We weren't trying to perform dark magic. We were working within known constructs and known limitations. Math and physics were on our side. So were foundational engineering principles. We just needed to devote time and resources to address each issue.

* * *

To create a starting point and a structure, I'd cleaved our new engineering efforts into two sides. Side one: radio frequency, the "RF" of the trade. Here was the interior circuitry where our modem would modulate or demodulate a "baseband" signal, meaning a signal in its original form that has not been modified to ride over a separate frequency.

The RF terrain inside our second-generation modem occupied only a few square inches of physical space, but it was a critical zone, a meeting ground between the physical layer, the baseband/RF world, and the logical—a data packet. Andy Borsa and Paul Nikolich, veterans of their craft, knew from long experience how a physical circuit board could affect electrical impulses and how layout adjustments needed to be made to compensate. If a circuit design called for a capacitance component value of 10 picofarad—the nomenclature for whisp-thin, sandwiched layers of metal and insulation—a designer might have to use a tiny, discrete capacitor component clocking in not at 10 picofarad but 9. The minute gap was required because the metal trace of the printed circuit board would itself account for the final whisper of capacitance. It was complicated terrain.

Then, the "digital" side of our RF architecture—the world of Bill Corley, conquering hero of the terrain. Here, we would leverage the power of digital signal processing, a field I'd first encountered during work at GE on the Comband project. My former GE colleague was working to tame the nastiness and ambiguity bedeviling cable television pipelines, battling assorted intruders, including micro reflections, ingress, and a host of other unwelcome aberrations.

These ingredients—our digital RF architecture, our data processing techniques, our signature Unilink-II protocol—would need to be encapsulated within the realm of an integrated circuit, a miniaturized successor to the gawky circuit boards of old. The plan was to present our coalesced brew of innovation, as if it were the recipe for a grand feast, to an unparalleled chip designer who was up for the challenge.

* * *

What we were trying to accomplish required an application-specific integrated circuit (ASIC) at a scale few had ever attempted. The integrated circuit initially came into being in the late 1950s, when the legendary Fairchild Semiconductor engineer Robert Noyce and a team of co-inventors devised the first working iteration, doing no less than inaugurating the modern computing age. Decades later, much of the work around ASICs had depended on the practice of schematic capture—physically drawing out the complex workings of electrical circuits before designing the pathways and the logic of the chipset itself. With the schematic capture approach to chip design, electronic circuits are painstakingly sketched out in mechanical, hard-copy style, with a designer mapping the inner workings and the intricate pathways. This design is then translated into a detailed set of building blocks that describe the circuitry of a chipset.

An experienced chip designer I was not, certainly not on this scale. But I knew enough to recognize schematic capture wasn't going to work. Not with the timetable we faced. Not with the scope of what we needed to create. And not without additional talent. I had hired John Ulm to begin to draw out the parameters of our Unilink-II protocol, but even John had never taken on a chip development project of this magnitude. What I also knew was this: A highly skilled and experienced chip developer at the top of their game could be counted on to produce somewhere around 300 lines of code per workday. Of that daily output, it was reasonable to figure we'd discover a dozen or so errors in the work requiring resolution or repair.

All of which made it remarkable when a young man named Kurt Baty showed up at our office in cowboy boots and a bright red shirt. He was an oddity from the start: a collector of rare international coins whose sideline hobby was hunting rabbits in the countryside of England with the aid of a trained raptor. During our first face-to-face meeting, he'd told me straight up that he ritually produced error-free code—an accomplishment I believed to be impossible. "Nope," he corrected me. "That's me. I'm that guy."

I'd contacted Baty originally on the advice of David Bither, a former Proteon colleague I'd remained in touch with. Bither and I had gotten along well in the Proteon days, partly because I'd given him wide leeway to do his job. He had gone on to work for Apollo Computer and also had done consulting work for Massachusetts-based Stratus Computer, which is where he'd come across Kurt Baty. Bither, a tell-it-like-it-is engineer, had come away impressed, telling me this pioneering chip designer was the right person for what I was attempting to accomplish. But as usual, there was a problem. Baty had zero interest in LANcity or my project. "I don't know who you are," he'd told me abruptly the first time we spoke over the phone. "Also: You can't afford me."

I persisted with a second phone call a week later, dangling a challenge that proved to be magnetic. The clincher was what I told Baty about our hours-long design review meeting with DEC, when Bill Hawe had walked out of the room, telling us we were trying to do something that could not be done. That had stopped Baty cold. I had piqued his interest by telling him a legendary DEC pioneer doubted he could do what we were trying to do. Baty had worked on projects with the big computer company before. He knew who Bill Hawe was.

A few weeks later, Baty showed up at the office, unannounced, and asked to see me. With zero formality or small talk, he laid out the terms of his engagement. He charged $1,500 per day. He would maintain his own work schedule. He demanded I outfit him with the latest high-powered Sun Microsystems workstation, a powerful machine that carried a six-figure price tag—well beyond our means. He insisted that he be in charge of LANcity's integrated circuit development team, with authority to demand work and enforce schedules.

It was a take-it-or-leave-it proposition. He had made it clear to me that he didn't need the work or the money. In his mid-30s, Baty was already very comfortable financially. "He was cocky and a character," Chris Grobicki recalled. "But also very driven and professional."

By now, timing was a big concern. We were staring down a DEC-imposed deadline of October 1992 to simulate a working chip. We needed help. And now, we had it. Within two weeks, the talkative Texan had rented an apartment nearby and was settling into an office cubicle.

Kurt Baty wasn't the only person who was becoming intrigued by our work. Some of our partners on the vendor side seemed to recognize something important was brewing. A friendly Sun Microsystems sales representative had arranged to loan me three expensive workstations for successive trial periods of 60 days, knowing I couldn't afford to buy them outright. Similarly, a Hewlett-Packard sales representative came up with a way to get us a pair of state-of-the-art, high-frequency spectrum and network analyzers that, at $250,000 each, LANcity could never afford. He carted the two pricey machines to my office on the promise that we'd return them as soon as we could, at no cost to us. Proof again that it helps to be nice to people.

I quickly realized Baty hadn't just been boasting about his personal productivity. Forget 300 lines of code with a dozen or so errors. To my astonishment, our newest team member was producing 1,500 lines of complex software instructions in one workday.

With no errors.

Zero.

At our fourth-floor office in Andover, we had ourselves a genius.

* * *

Kurt Baty had always been gifted. As an elementary school student, he dominated the "Midget" division of the Cedar Rapids, Iowa, Chess Club in the early 1970s. He had learned to play as a four-year-old, leaning over a board across from his father, a US Army Signal Corps engineer who built optical targeting systems during WWII. The younger Baty remembers his father cornering him into checkmate four or five times. That was that. He never lost again to his dad.

Or to hardly any other opponents. A few years later, after a teenager outmaneuvered the seven-year-old Baty in a junior chess tournament, the young apprentice obsessively pored over the match, committing every move to memory. When he met the same rival again, Baty methodically tormented his opponent, picking off chess piece after chess piece before the rival player finally surrendered. "It was like pulling the wings off a fly," Baty recalled. "I was a vindictive little shit."

He tore through advanced physics and mathematics courses in high school on the way to an electrical engineering degree at the University of Iowa, followed by an advanced degree from Iowa State University. Tech companies came calling. Baty accepted an offer from the up-and-coming mini-computer maker Data General Corp., whose pioneering work would be canonized in author Tracy Kidder's book, *The Soul of a New Machine*. Soon after, the computer maker Stratus Computing lured Baty away. At Stratus, Baty worked in the emerging field of advanced microprocessor design—taking complex arrangements of logic and mathematics and imposing them onto silicon surfaces that would become the nerve centers of a new computing era. Baty would go on to design more than 50 special-purpose integrated circuits—including ours.

Baty made two important discoveries while at Stratus: One, he was spectacularly adept in a field that was about to explode in terms of demand and activity. Two, he could make a lot of money because of it. Royalties tied to chips Baty had built for Stratus had made him a millionaire by his mid-thirties. Sensing an inflection point was at hand, he launched his career as an independent chip designer in 1989. His business card confidently stated his purpose: "Design challenges and problems solved."

The tagline resonated loudly and clearly to me. At LANcity, we certainly faced our share of challenges. And yes, we needed help solving them. I'd been able to recruit Baty based partly on the pledge of a generous compensation deal—$1,500 per day was a big number for us—and an evergreen royalty arrangement that would give Baty a share of the action for any LANcity cable modems that contained the chipset he had designed.

The goal was a lofty one. From our office on Brickstone Square, we were trying to take the workings of four clipboard-sized circuit boards decorated with hundreds of tiny electrical components and squish much of their functionality onto a pair of six-inch by nine-inch boards that would become the nerve center of a new cable modem. From there, it was Baty's job to find a way to take the Unilink-II protocol and translate it to a chip that would then be inserted into one of our modem's shrunken boards. The sheer numbers were daunting: Our chip would contain more than 155,000 logic gates, the on/off switches of digital data processing. The number of computations that would have to be carried out every second would top four billion.

To say our new friend Kurt Baty was an iconoclast would be a grand understatement. In the heat of the moment, he would sometimes pack up his leather briefcase and take off, providing little warning other than to tell me he'd be back . . . sometime. Off he would go: to scope out rare coins in the Soviet Union or to clamber about the English countryside. "You have me for another three hours, Rouzbeh," he would tell me. "I don't know when I'll be back." On other days, he'd arrive in the morning and leave almost immediately after discovering the work he'd instructed our developers to accomplish the night before hadn't been done or had not accomplished what he'd wanted it to do. Sometimes, I'd lose track of him, only to discover him down the hallway shooting the breeze with the guys in our manufacturing bay. But I didn't flinch. I knew that in Kurt Baty, I'd come across a rare breed. "Kurt was absolutely brilliant," John Ulm recalled more than 30 years later. "I mean: I'm good. But Kurt was absolutely one step above me."

At least once a week, Baty would call me over excitedly to show me the favorable results of the latest test or the accomplishment of a major milestone. There were days when we'd run torrents of signal waves across our powerful Sun Microsystems workstations. Hovering around the workstations, Baty would point with pride to the glowing readouts appearing on the screen. "See that?" Baty would ask me. "You are one lucky man, Rouzbeh. I did this. I made this work."

One of the first things we learned about Baty was his insistence on doing things incrementally and making sure that, once scripted, they worked. The philosophy made for an iterative design process: We'd complete the design of one discreet part of the puzzle, and then immediately, usually starting that same evening, put the work through a rigorous regression testing process. It was an ahead-of-its-time implementation of what would later come to be known as "agile" design—what you code today, you test today. This process took time because it chopped our work into hundreds of unique moments. But it was essential. Running tests on a broader set of functionality, only to find that we'd screwed up early in the going, would have required unwinding days or weeks of work to identify our faults and fix them.

I also appreciated Baty's insistence on moving forward. I'd worked with developers who were never quite satisfied with the results of testing. Left to their own, they'd prefer to test indefinitely, tweaking and adjusting in pursuit of some unattainable ideal of perfection. Not Baty. If we got to an acceptable point on a module or a milestone, we'd move on. But not without building in some safeguards. Baty insisted on flexibility throughout the design process, an approach that would later prove to be critical for LANcity and the entire cable industry. If my team suggested we use an 8-bit counter, Baty would design an upgrade to 16 or 32 bits. He insisted that every input within our chip be upgradable with software. Not one was hard-coded. That way, everything could be touched by software and corrected if needed. It would prove to be a smart, game-saving call in just a few months.

The timing was tight, leaving no room for mistakes. Our prototype chip had to work correctly the first time, hewing to a looming production deadline for the real thing. Baty realized what we were trying to do was too complex and too urgent for schematic capture. We would end up with enough mapped-out circuits to fill a football field with paper. Maybe two. Even a modestly complex design involving tens of thousands of gates might require dozens of hardware designers, take months to enter the schematics, and then be prone to mistakes requiring an elongated debugging cycle.

Instead, Baty convinced me a new process called simulation was the way to go. We'd use an approach known as hardware design language, specifically the Verilog tool, to model our chip. Thereafter, we'd use a software compiler to convert our model to a gate design. Using state-of-the-art simulation and synthesis tools would allow us to identify problems earlier in the design cycle, well before we went into production.

Still, I was nervous. We all were. There is a point at which every one of the millions of instructions you've written will be translated into a physical design. Even if we'd done grade A work, even after performing exacting simulation, there was still the possibility our circuits would have serious flaws.

Placing lots of components on a miniaturized surface allows for many chances of error. Baty frequently warned us that the Verilog tool worked both ways: It greatly sped up the design process in the hands of a master, but it also allowed a less-experienced designer to create garbage ten times faster. Even though we'd get back a handful of prototype chips, to-the-minute production schedules and compressed time demands within our DEC agreement were such that we'd have only a narrow window between acceptance of the first few test chips and authorizing the full production run that followed. Miss the window because of intensive debugging requirements or other adjustments to the original design and we'd be set back months.

Already, even before handing over our fabrication tape to the manufacturer, we were racing. A critical payment from DEC, scheduled for October, was tied to our success in creating the new silicon. There was no second chance. Either we met the specification or the money flow stopped and our dream of high-speed data over the residential cable infrastructure vanished. For these reasons, I was grateful Baty had insisted from the start on an iterative, prove-it-and-move-on design process.

There was one more Kurt Baty dictum: If open-source code was finished, proven, and available, the rule was: grab it. "God gave you two eyes . . . so you could plagiarize," Baty would chant from his cubicle. He was exaggerating. But the point was clear to everybody. Don't reinvent the wheel. He would draw from a vast reservoir of intellectual property maintained by Synopsys and its "DesignWare" repository, a library of proven work that could be incorporated into chip logic.

With time dwindling, we were closing in on completion of our pattern generation tape that would be sent the next week to our chip fabricator to be translated into a prototype chip. It had been a long journey. Our team's QPSK work alone traced back nearly six years. As Baty put the finishing touches on our tape, my heart pounded, and my already-poor sleep habits worsened. My vision was about to come alive in the form factor of a new, agile, smaller, sleeker modem that would prove data could flow flawlessly over the residential cable infrastructure. If it worked.

* * *

On a Tuesday morning in December 1992, a Federal Express courier making the rounds of Boston-area tech companies dropped off a package. Inside was the heartbeat of our second-generation cable modem, a key to our broadband vision. I peeled back a protective sheath to find a handful of miniaturized integrated circuits, gateways to a long-imagined future. These innocent-seeming rectangles would determine whether LANcity succeeded in transforming the nation's cable television infrastructure or disappeared into the start-up cemetery after failing to meet a technical milestone enshrined in our deal with DEC. A payment of $557,000 from our partners at DEC hung in the balance.

I handed the new chips to my colleague Mike Sperry, who set off on an hour-long drive to Westborough, Massachusetts, where earlier we had handed our contracted hardware manufacturer, Design Circuits Inc., a set of customized, printed circuit boards. To buy time, we had designed and assembled these intricately architected boards in parallel with our chip development work, leaving an empty position near the center of a green surface where the new chips would nestle. Up close, the new circuit boards for our second-generation modem looked like miniature cities: neatly arranged streets and pathways leading to a succession of vibrant colors and geometric shapes; inductors that looked like miniature doughnuts wrapped in thin red wires, edging up against a neighborhood of tan-colored, cylindrical capacitors; and rectangular multiplexers with neat, dark-gray edges sharing space with pin-like diodes decorated by colorful end caps. I was mesmerized by the intricacy we'd created and the connectedness of it all; how each tiny component played its own vital role in governing the flow of electrical currents. Remove any of the hundreds of piece parts and the whole thing collapses.

Each chip was equally meticulous from a design standpoint, sprouting 160 conductive pins that connected to corresponding slots. Because of their physical intricacy, it took three days to position the chips delicately. Mike kept in close touch with our manufacturing partner, monitoring the progress like a nervous parent as the chips were carefully put into place. Three days later, he drove back to Westborough, picked up the newly altered boards, and returned them back to our office. We'd asked DCI to populate three sets of circuit boards, reserving a pair of our sample chips untouched. Just in case.

LANcity's chief software architect, Gerry White, had been running our software code through a demanding testing regimen that cobbled together powerful hardware and protocol tools to simulate the workings of a real-world IP data network. Gerry had done everything possible to create a virtual simulation of our hardware, a sort of invented, parallel universe we'd devised in order to maximize our time window for testing. Having doggedly poked and prodded at this software architecture for months with a sense of determined insistence, by now Gerry, along with team members Ed O'Connell, Cindy Mazza, and Paul Chamberlain, were satisfied that our simulated test bed was ready to do its work once our new silicon was in hand. The moment of truth was looming as we prepared to marry our pretested software stack with its companion hardware for the first time. It would be a nerve-wracking, nearly month-long regimen of diagnostic tests during which I could barely sleep.

Our tests were happening atop a makeshift testing table. Tangles of wires and cables, circuit boards exposed out in the open, power cords jutting from outlets, developers standing, peering over one another's shoulders, many of us unable to sit for long out of nervousness. The scene reminded me of the labor room in a hospital maternity ward.

The first test was an elemental trial: Could our chips survive the inaugural tickle of an electrical current? With our boards plugged in and powered, half a dozen team members stood nearby, watching nervously as we powered up the boards by running current through our circuits, relieved to find no evidence of an errant path—the dreaded short circuit. One "go/no go" test down, three more to come.

Stage two was all about starting a dialogue with our chips. We needed to toggle each of the logic gates in our chipset one by one. We were probing at a foundational level to affirm the logic elements of our Unilink-II protocol were working. Increment by increment, our chips were replying with a positive answer. We'd made it past stage two.

Next, a set of critical "loopback" tests. We knew the logic gates packed into our chipset were working. The question now was whether we could send larger blocks of data running through these data layers, with the input to the transmit chain perfectly matching the output from the receiver chain—the basis for the term "loopback." Here, we were still operating within the isolated chipset environment itself—no data packets were yet traveling outside of the silicon—but the results of our loopback tests were positive. Threads of data were moving around without incident within our integrated circuits. We repeated the tests day after day until we'd gone through all of the assigned blocks.

It took nearly three weeks to affirm that the basic plumbing of our silicon-powered Unilink-II protocol seemed to be working. Next, we had to determine if the simulation work that Gerry White's team had been conducting—emulating the workings of our modem without having the physical hardware in hand—was real. We needed to move beyond merely conversing with our chips and instead test our full cable modem architecture—a meeting ground of the analog with the digital, of old-school components paired with new-age silicon.

Two especially worrisome questions ran in my mind all along. Our software had performed ably within Gerry's test environment. But could it work within the new cable modem hardware, over the lone megabyte of memory our compact design provided? Because of size and architectural constraints, we were relying on a compact but powerful flash memory chip produced by the chipmaker Advanced Micro Devices of Santa Clara, California. Its maximum capacity was one megabyte. That's all we had room for.

Once again, relief. We passed a major hurdle. Our software had been loaded to the circuit board and was responding. That left the last big question: Could we pass data packets through our modem? As in, in one side, out the other—a microcosm of the way IP data would flow across cable television lines in the real world. In other words, does the modem we've spent years of our lives on actually . . . work? We'd spent nearly a month doing everything to our modem except transmitting packets—the basic premise of TCP/IP protocol. Now, it was time.

It began with a "send" command from a keyboard attached to one of our Sun workstations. Cindy Mazza had a front-row seat in the action. One of the 13 original employees of LANcity, Cindy was a determined, diligent software developer, seemingly always a step ahead. What Cindy was not, however, was the most vocal member of our team. She was our quiet warrior. Her most common auditory contribution was the steady tapping of a keyboard. It was surprising, then, that the first thing I heard was a vowel-intensive yelp of pure, unbridled, extemporaneous joy, straight from the throat of Cindy Mazza. No Celtics fan could have ever delivered a more convincing whoop of victory. She had been sitting in front of a workstation monitor, six or seven of us hunched closely nearby. Because of this, she enjoyed a close-up view of what had just happened during a moment that will live forever in my heart.

In an instant, there it was: the digital fingerprint of a single, lonely data packet flashing on a connected monitor. The small speck of data had passed through our software and our new silicon. One tiny, isolated, measly packet, a drop in the digital ocean. But it was all we needed to see. It had traveled unblemished, traversing not just the interior of a new siliconized cable modem but across time itself: Years of intense, demanding work powered by unrelenting conviction. The outburst rang out like a shot, expressing aloud the greatest "wow" moment of my professional life.

A final lab trial followed as we tested our cable modem over the mock cable system we'd installed in our own lab. Knowing that digitized video content presented challenges for data networks because of the density of data packets, Chris Grobicki and I arranged to use a brief video clip of a flying dove, stored on a CD-ROM disc, as a proof point before arranging a handful of real-world modem implementations. Huddling in a cubicle, Chris and I waited nervously to see if our dove would fly over our in-house cable system. In a flash, there it was: our broadband bird, feathered wings flapping, the video stream shooting across our amplified run of cable into our LCB modem through the innards of a connected computer and, now, onto a screen in high-resolution splendor. We could see individual feathers bristling. The image was gorgeous, a kiss from an angel.

Within days, we were flooding our system with waterfalls of data, watching like proud parents as bitstreams flowed without interruption from the powerful packet generator. Our new cable modem—software, new silicon, our new RF tuner, our Unilink-II protocol, the entire stack—was working. We'd earned another cash injection from our partner DEC.

By now, we had done enough testing to know we could confidently say "go" to our volume chip-manufacturing schedule. Thankfully so: Our relationship with DEC, and the funding behind it, depended on signing our first paid customer for our next-generation modem by August. Now, with our dove flying and our tests conclusive, we had the confidence to pivot immediately to field tests over several working cable systems.

We also staged a few personalized demonstrations. Working with DEC, we rigged up the home office of a professor at the Massachusetts Institute of Technology who had expressed reservations about whether our technology would work. We'd connected our new modem to the cable system that served his neighborhood and invited him to download to his computer a high-resolution video displaying the launch of a NASA space shuttle. Accustomed to the slow grind of low-speed modems and phone lines, our friend from MIT was about to leave to take a shower while he waited for the file download to complete. But as he glanced back at the monitor, he discovered the video was already on his screen: proof of our broadband vision, playing on an infinite loop.

CHAPTER 7

DEATH STAR

A perfect storm was taking place behind John Malone's 500-channel dream.

Cable television was on a roll. By the early 1990s, more than 50 million US homes were connected to cable television—nearly 60 percent of the country. For an industry that sprang to life as an oversized version of a TV antenna, the achievements were undeniably impressive.

Cable channels that were once dismissed as irrelevant by the likes of ABC and CBS started siphoning off viewers. In the all-important primetime environment—evenings spent at home watching television—the percentage of the audience attracted by the Big 3 broadcast TV networks had plummeted to 62 percent, down sharply from 85 percent in 1980. Cable TV was changing what Americans watched on TV and what they knew about the world. The CNN journalist Bernard Shaw's dramatic, you-are-there reporting from Baghdad during the 1991 onset of the Gulf War had reinforced cable as a power player on the media scene.

On the ground, there was little standing in cable's way. Almost without exception, a one-to-a-market rule held: The company that possessed the legal right to wire a particular town or city rarely faced direct competition from another cable company. It was an investment dream: Little head-on competition, coupled with the freedom to set prices with little or no pushback.

And for that, the industry could thank the federal government. It had been several years since the passage of landmark legislation—the Cable Communications Policy Act of 1984—that freed much of the cable industry from rate regulation. Since then, it had been possible for cable companies in most communities to set their rates with little local government oversight. The result was a surge in revenue and cash flow, much of it plowed back into channel expansions and support for new TV networks, creating a virtuous loop that made the cable industry the envy of Hollywood moguls and legacy TV executives.

On the flip side, it was the bane of consumers.

By the late 1980s, a multiyear succession of annual rate increases began to provoke the wrong kind of attention in the nation's capital. Legislators and their aides were hearing an earful from constituents who were growing weary of getting socked with cable TV rate hikes every January. A 1992 survey conducted for the trade magazine *Cable World* found the percentage of cable TV subscribers complaining their rates were "way too high" had shot to 37 percent from 17 percent three years earlier. Although much of the industry operated with relative restraint, exorbitant rate hikes imposed by a few companies—"bad actors" in the industry's lobbying parlance—had darkened the whole industry's persona.

As a result, serious efforts had been gaining ground among key members of Congress to rein in an industry that looked to be abusing its pricing liberation. Led by Democratic senators Al Gore (Tennessee), Howard Metzenbaum (Ohio), and Daniel Inouye (Hawaii), along with US representative Edward Markey (Massachusetts), a move to re-regulate cable TV rates was gaining steam as a populist issue that transcended party lines. Lawmakers seeking reelection, or merely looking to score points, knew they had a winning issue here: socking it to the cable guy.

James Mooney, the president of the cable industry's main advocacy organization, NCTA, had been telling his board of directors that staving off federal rate legislation was by now likely impossible. Mooney advised working with legislators and their staff to try to soften the edges of the bill they were working to draft. He told *Multichannel News*, "The cable industry has got to come to grips with the fact that rate increases in excess of the overall inflation rate are making people very angry, and are now almost entirely the sum and substance of the cable industry's political problems in the Congress."

Mooney was a politically savvy DC operative with deep Beltway connections and a signature brown leather cigarette case reliably nearby. Employing an almost Shakespearean eloquence, Mooney had become an industry hero for championing the 1984 federal law that unshackled the cable industry from local rate regulation in the first place. Now, his reading of the tea leaves was correct: The 1992 Cable Act passed both chambers of Congress and, on the eve of Congress's 1992 adjournment, won enough support to override a presidential veto from George H. W. Bush.

Congress had expressed its intention to rein in cable TV rates. Just how that happened would be up to the FCC. Cable industry executives braced for impact. The agency's new chairman was a Washington, DC, attorney named Reed Hundt, a longtime acquaintance of the future vice president, Gore. Hundt would rise up as chief antagonist to a cable industry that watched in dismay as the formulas the FCC put forth for setting a ceiling on rates would prove to be more onerous than even the cable industry's worst-case scenarios had foreseen. TCI's John Malone, in a 1994 *Wired* magazine interview, would famously (and regrettably) remark to a journalist that salvation for the cable industry would depend on somebody doing away with Hundt. "All we need is a little help," Malone was quoted as saying. "You know: shoot Hundt! Don't let him do any more damage." Malone later apologized to Hundt for the untoward remark, and TCI's press department issued a public statement walking back the comment as "obviously not a serious one."

Even so, Malone's broader point was telling. The FCC, under Hundt, had badly wounded the cable industry. Government-mandated rollbacks of some cable rates below the level they were at in the autumn of 1992 caused carnage for highly leveraged cable companies that suddenly struggled to make their numbers work. Swooning stock prices and the undoing of major industry transactions—including one stunner, the proposed merger of TCI with the regional telephone company Bell Atlantic—told the tale of an industry in turmoil. TCI blamed the FCC, which had imposed still another rate rollback early in 1994, for the collapse of a proposed deal that one journalist, the *Multichannel News* writer John Higgins, had labeled "the defining moment for the building of the information superhighway." The FCC, in the view of one industry financial analyst, Tom Donatelli of Wertheim Schroder & Co., had undertaken a campaign of "regulatory terrorism" on the cable industry.

But restrictive rate regulation was just one of the two macro pressures that would send the cable industry into a tailspin. Another major threat came from outer space.

For the last few years, there had been serious exploration about using a different sort of technology to deliver "cable television" channels. Industry watchers called it DBS for "direct broadcast satellite." Regular people would come to call it by a more familiar name: DirecTV.

The name was chosen for a reason. With satellite television, signals would hail down from a satellite hovering in space to a pizza-sized receiving dish—no intermediating cables necessary. This "direct" pathway to the customer meant virtuous outcomes for the costs of maintaining the system, including fewer visits by technicians, or "truck rolls" in industry parlance.

DirecTV was the brainchild of engineers and executives working in the Los Angeles seaside suburb of El Segundo for a General Motors subsidiary, Hughes Aircraft Co. The company specializing in satellite communications was one of the country's prominent military contractors. Hughes Aircraft's satellite TV effort was led by a determined, confidence-exuding executive whose carefully trimmed white beard and urbane manner suggested the jolly mien of Santa Claus more than that of a fierce corporate warrior. But Eddy Hartenstein was convinced the cable industry, with its spiraling prices and seemingly flippant attitude toward customer service, was ripe for attack.

There were others who had come to the same realization. A low-key but determined television broadcaster from Minnesota, Stanley Hubbard, was also designing a multichannel DBS service. In Colorado, a twang-tongued entrepreneur from Tennessee had like-minded ideas, betting that the nation was eager to install satellite dishes on rooftops and yank the cable. Charles Ergen had until now made his living peddling giant, backyard satellite receivers that could capture signals from cable channels like HBO. He would go on to build a satellite TV service, Dish TV, that at its height would boast more than 15 million subscribers and become a persistent thorn in the cable industry's side. Even some prominent cable companies, sensing the possibility of a better mousetrap, tried to carve out a role in the DBS business. Charles Dolan's Cablevision Systems launch a short-lived satellite TV service, Voom, in the autumn of 1993. Two years earlier, a separate cable industry consortium had tried its hand with a like-minded DBS service targeting rural households: Primestar.

But it was DirecTV, with its promise of lots of channels, attractive prices, and the delight, at least for some weary customers, of kicking the cable guy in the shins, which loomed as the top threat to the cable television industry. A September 1993 issue of the trade magazine *Cablevision* featured a smiling Hartenstein on the cover, adjacent to a foreboding headline: "Death Star: Will DirecTV's DBS service give cable operators a run for their money?" The answer, it would soon become evident, was "yes." DirecTV was determined to beat cable television at its own game.

* * *

As Hughes Aircraft Co. began to plot its plan for satellite television, cable industry sentiment was curiously mixed. Some nonbelievers, in fact, joked that the acronym of DBS should stand for "don't be silly." The failure of an earlier, much-ballyhooed satellite TV venture backed by Washington, DC, company Communications Satellite Corp. may have been one reason. "It was such a bust," recalled Char Beales, a well-known cable industry marketing figure, "that it lured a lot of operators into complacency when DirecTV launched."

Beales began her career as a TV station marketing employee in the early 1980s, later accepting a job with the Washington, DC, cable industry trade association NCTA. In 1992, Beales began a long run as the leader of a national cable industry marketing association called CTAM, with the acronym originally standing for Cable Television Administration & Marketing Society.

Beales brought a tell-it-like-it-is mentality to her job. Almost from the start, she had warned anybody who would listen that DirecTV was to be taken seriously. But it took a while for her industry peers to adjust to the new reality. In 1993, CTAM's annual conference reflected a fairly benign viewpoint, presented under the theme "The Customer Speaks." But two years later, as DirecTV's subscriber count swelled past one million, the message was more strident. "By the time we got to San Francisco in 1995, it was: Holy shit," Beales recalled. "The theme was, 'Wake Up and Smell the Competition.'"

Like Beales, TCI's hard-charging CEO John Malone took satellite TV seriously. An engineer early in his career, Malone recognized that DirecTV could get there first with a new type of "digital" television signal—video streams cloaked in the language of computing.

Over the next few years, digital video technology would emerge, albeit indirectly, as an important enabling element for our work at LANcity. Besides the promise of more channels, an early propellent was high-definition television, a new kid on the TV block that would enliven TV screens with improved visual resolution, bringing more lifelike pictures to the screen.

In the late 1980s, cable companies had become deeply concerned about where HDTV was headed. Their fear was that the adoption of an HDTV technology called MUSE, which was being nurtured by a consortium of Japanese companies, could wreak havoc on cable's ability to squeeze bandwidth-hungry HD signals down cable's crowded analog pipes. If the Japanese system became the standard by which TV sets displayed signals, displacing the National Television Standards Committee scripture that had long ruled in the United States, it could spell economic catastrophe for a cable industry that simply didn't have room to push out these expansive signals.

Cable needed a hero. And here he came: A bespectacled TCI lieutenant named John Sie began making the rounds of policymakers and peers, advising anyone who would listen to be wary. Sie insisted there were better, nimbler HD technologies that would not require a retooling of the US television system at large. In fact, he pointed out that by using a different breed of digital technologies, cable operators could accomplish the reverse of what the MUSE system proposed: delivering more channels, not fewer, across the same spectrum. A demanding, driven executive who had a habit of phoning subordinates at home in the early-morning hours, Sie proved indefatigable. His persistence in championing alternative technology—with industry colleagues, with technology companies, with journalists, and importantly, with standards bodies and with the FCC—began to shift the conversation. In 1990, the American National Standards Institute issued a rare about-face, reversing its earlier decision to support Japan's MUSE system.

But Sie's success in derailing MUSE would have consequences elsewhere. The fervor burbling around HDTV and digital television technology had caught the attention of Hartenstein and DirecTV. Even before there was a final standard for digital television in the United States, DirecTV made a key decision. It would launch its satellite television service using a new digital technology, one that adhered to specifications being considered by an international body called the Motion Picture Experts Group (or MPEG). Early adoption of the MPEG scheme, untested at scale, was a big risk. But if it worked, DirecTV could shower the world with more raw television tonnage—hundreds of channels raining down to rooftops—faster than anyone.

It worked.

The first TV commercials promoting DirecTV smacked the cable industry across its forehead. "What if I told you everything you know about TV is about to change?" asked a narrator, as logos of dozens of popular cable channels appeared on the screen. DirecTV had beaten cable to the digital television punch, proving that a game-changing concept known as signal "compression" could do exactly what it promised: squeeze more channels into assigned frequencies. If Americans wanted more TV channels, DirecTV was the way to get them.

Video compression was about arithmetic. Any given scene of a television show is a sequence of individual frames. And every frame is a mashup of tiny nuggets of information describing color and light. The revelation behind video compression was that, often, these pixels remained constant, frame after frame: Think of the glowing lampshade atop the desk in a private detective's office within a murder-mystery series. On a TV screen, the lampshade is the exact same compilation of pixels in every frame where it appears. Thus, it was possible to capture the information needed to convey that image only once—and borrow it for subsequent frames. Doing so shaved the bandwidth requirements of video so much that DirecTV could pack multiple digital channels into the satellite transmission capacity formerly required for just one analog channel, which is exactly what DirecTV did.

John Malone had surveyed the shifting landscape. He knew digital technology was game-changing. He knew cable needed to formulate a digital strategy to fend off the satellite TV threat. At the cable industry's December 1992 Western Cable Show in Anaheim, California, Malone met with journalists to talk about what TCI was going to do.

It was the press conference heard 'round the world. Malone told reporters that thanks to digital compression, the cable industry was gearing up to deliver as many as 500 channels to its customers—along with futuristic applications like on-demand television (watch movies and TV shows whenever you want to) and on-screen pizza orders.

"Television," said Malone, "will never be the same."

Nor would expectations for the cable industry. The 500-channel reference made its way into the lead paragraph of an article the next day in the *New York Times*. With a single statement, John Malone encapsulated the cable industry's strategy of attack. Cable was going to fend off the emerging satellite TV category by focusing efforts mostly around a familiar business: television.

I watched the developments with keen interest—and, perhaps ironically, a sense of hopefulness. True, we were about connecting to computers, not TV sets. We had zero involvement in the television side of cable television. And yet, I was excited to see what was happening for one reason: bandwidth. Until now, cable systems had been stuffed to the brim with analog television channels, each occupying the prescribed 6 MHz. Now, the industry was out to prove that the same 6 MHz swath of bandwidth could accommodate four or more digitized television channels. That meant providers could pack more channels into their systems and still leave room to experiment with other services. I flashed back to my GE days. All bandwidth was good bandwidth.

A second benevolent impact of cable's digital conversion had to do with two-way signaling. The combination of digital video delivery and bidirectional pipes could enable cable companies to devote some of their newfound capacity to presenting individual movies and TV shows "on demand"—available to be watched whenever a user pressed a button on a remote control. Sending movies to the home for per-viewing fees had long been a fever dream of Hollywood, tracing to 1953 when a company named International Telemeter Corporation bolted coin boxes onto the sides of TV sets in Palm Springs, California. A payment of $1.25 gave viewers the chance to watch the Hollywood ingenue Ginger Rogers in the movie *Forever Female* from the comfort of their couch. Digital video technology was taking this same idea into new territory, presenting movie fans with an alternative to the cumbersome Blockbuster Video experience.

Again, here was good news for LANcity. A key to stealing away the movie rental business from the likes of Blockbuster was making cable systems bidirectional—allowing for upstream signaling that would enable customers to issue "buy now" and "press play" commands. As we knew from our work in Stow, functioning bidirectional systems were a must for delivering broadband data over cable. Now, Hollywood was on our side.

A separate advantage for us was an inescapable physical reality: satellites hovering more than 22,000 miles in space weren't optimal for delivering high-speed data without introducing significant delay. But on-the-ground cable systems were a different story. They could establish a competitive advantage over satellite TV by making high-speed data real.

A perfect storm was taking place behind John Malone's 500-channel dream. Digital video compression was freeing up bandwidth. Video-on-demand aspirations were encouraging cable companies to make two-way systems real. And the threat from DirecTV would impress on cable companies the need to invest in new services.

Then, a critical piece of the techno puzzle. The cable guys were deep into an engineering makeover that would clear the pathway for the modern broadband revolution. It would work its magic by seizing the power of an elemental component of the natural world: light.

* * *

Growing up in Houston, Texas, where he rattled around in a late-1970s Chevrolet Vega, Rick Guerrero had always been a math kid, captivated by the sureness and certainty of numbers, ratios, and formulas. That, plus a sensibility for geography. Guerrero had learned how to draw detailed telecommunications engineering maps in a night class at the University of Houston—a skill that would come in handy when a local cable company, Bellaire Cable TV, hired Guerrero on contract as a system designer in 1979. He made enough money to devote himself to his new calling, spending the rest of his career designing cable systems for industry standard-bearers like Times Mirror Cable and Cox Communications. In doing so, Guerrero would have a front-row view of a technology transition that would usher in the information superhighway: fiber optics.

Fiber optics had been on the telecommunications scene for a long time. In 1965, an ITT engineer named Charles Kao proved that pristine glass fiber could be used as a medium for transmitting light waves on which information had been imprinted. Telecom companies, including AT&T, GTE, and the long-distance carrier MCI, followed in the 1970s with field trials and implementations, reasoning that fiber lines were well-suited for long-distance transmission of digital data that increasingly was coursing over the land. The benefit: Light did not require amplification. Lasers situated at one end of the network could emit information across a fiber connection to a distant receiving node with no electronic amplification, and almost no loss of fidelity. Fiber-optic networks could wipe out huge numbers of amplifiers, potentially easing one of the big technical impediments to broadband data delivery over cable.

Cable companies had begun to (literally) see the light as early as the mid-1980s. In Tampa, Florida, the Denver-based cable company Jones Intercable was among the first to replace coaxial cable trunk lines with fiber lines, displacing amplifiers and yielding big improvements in signal quality. Cable's fiber fervor also drew oxygen from the efforts of Louis Williamson, a telecommunications engineer who was early to the game of marrying fiber optics with traditional coaxial cable lines. Williamson grew up in Norfolk, Virginia, displaying a youthful fondness for disassembling any nearby appliance: a radio one day, a backyard washing machine another. As a student at Virginia Tech University, he had taken a few classes devoted to fiber optics, examining how lasers could express electronic pulses across the medium's thin strands with tremendous fidelity.

After landing an engineering job with Time Warner Cable's predecessor company, American Television & Communications (or ATC), Williamson's interest deepened as he learned about headaches involving lengthy cable amplifier cascades. In experimenting with fiber optics, ATC was devising the early iterations of a new network design that would soon take over the cable industry at large. It involved smallish, neighborhood-serving areas—"nodes" in industry lingo—that would vastly improve spectral efficiency. With nodes, cable's fiber connections would be shared by only a few hundred customers, not tens of thousands, as in the old days of amplifier-strewn trunk lines. Each node effectively operated as its own miniaturized cable system, making for a scenario where users had plenty of available bandwidth at their disposal.

But acclimating to fiber optics wasn't easy. Technicians underwent intensive training. "Imagine 144 fibers flopping in the wind, and you're trying to match the colors from a sheet somebody handed you at the office," explained Guerrero. Beyond the technical challenges, cable companies had to overcome skeptical customers as they built out their fiber-meets-coaxial networks. In the town of Santa Clarita, California, homeowners lined up at a city council meeting until 1 a.m. to lambaste the local cable company MediaOne Broadband, which was using a faux stone material to disguise its new fiber cabinets. "They didn't like the way the fake rocks that we were putting over our boxes looked in the streets," recalled Teresa Elder, at the time a regional vice president for MediaOne.

But fiber optics and nodal architectures were too promising to ignore. Cable company executives realized they had a unique advantage in the race to fiberize the nation at the neighborhood level, and they were determined to harness it. The coaxial cables that already snaked their way to millions of addresses represented the industry's secret sauce: a sturdy, high-capacity, already-in-place "last-mile" path to the home.

The idea, then, was to pivot to a hybrid sort of network. Fiber lines would convey TV channels at nearly the speed of light—and with minimal signal loss—to neighborhood nodes. There, they would connect to the existing coaxial feeder network, running over cascades of only a few amplifiers—not 40 or 50. With a hybrid fiber-coaxial (HFC) network, operators could break one large system into lots of smaller chunks, unleashing more capacity for individual users. In Alexandria, Virginia, for example, Jones Intercable busted its cable system into 461 separate optical "nodes" by the fall of 1995, setting the stage for the introduction of services like video-on-demand, cable telephony, and ultimately, broadband data delivery.

If digital video and fiber optics took off the way we hoped they would, lots of amplifiers would be retired, systems would get compartmentalized, available bandwidth would soar, the two-way cable infrastructure could finally blossom, and the overall performance and reliability of cable systems would vastly improve. Our dream of broadband data over cable looked to have a legitimate chance of happening.

What could possibly go wrong?

* * *

Fallout from the 1992 cable regulation law hit the cable industry hard. Under the new legislation, cities and towns with the will and the staffing to monitor their local cable company could decide whether to permit requested rate hikes. Those without could punt to the FCC, which had set in place a complicated cascade of formulas designed to manage rate increases or the lack thereof.

As the law's provisions began to ripple through the ecosystem, the pain became evident. The economic elixir of raising rates for cable TV service had all but vanished, putting pressure on the industry's all-important ability to generate cash. That cash was paid not just for running a cable company but also for paying back banks, insurance companies, and other lenders that tended to be unyielding about enforcing their lending covenants. Estimates published by *Cable World* magazine on April 5, 1993, suggested that because of the new cable rate regulations, the domestic cable industry's yearly cash flow total could sink by as much as $975 million, a nearly 10 percent slide. A cable industry stock index maintained by the magazine showed a 15 percent overall drop in share values from March 31 to April 8. One of cable's patriarchs, Continental Cablevision cofounder Amos Hostetter, projected tough sledding ahead, saying the second quarter of 1993 would be "the hardest three months in my 30 years in the cable industry."

If that wasn't enough, an internal power shift was also afoot as economic clout began to drift away from cable operating companies and toward the national programming networks that populated cable channel lineups. Once-obscure newcomers to the US television scene—the likes of The Discovery Channel, CNN, Nickelodeon, HGTV, and dozens of others—had by now become familiar household brands, prized entrants on the American media menu. And they knew it. Thus, as the impact of rate regulations rippled, an increasingly common scenario was playing out within the cable business: It was starting to get very expensive for cable companies to offer cable TV channels to their customers.

"Suddenly you were in a conundrum," recalled Patty McCaskill, who negotiated programming carriage contracts throughout the 1990s and later for the cable company Charter Communications. "Do I give them more money and keep them, or do I drop them and anger my customers, who had their cable bills justified because they had new channels that they had gotten used to?" Most of the time, the answer was the former: give them more money.

Other TV programmers piled on. A provision in the 1992 Cable Act meant that for the first time, local, over-the-air TV station owners could cash in by demanding that cable companies pay them in the same way that they paid cable-original channels like Discovery or The Learning Channel. Broadcasters were armed with a newfound power derived from an FCC prescription known as "retransmission consent," which effectively meant cable companies would have to negotiate for the right to send out over-the-air TV stations through their pipes. It was a tough time. The combination of rate regulation, more and higher programming costs, and the satellite TV crusade made for an unwelcome convergence.

To regain momentum, cable companies tried a grab bag of new tricks. Some industry executives had high hopes for an emerging delivery system known as "near video on demand." The clumsily named predecessor of (real) video-on-demand allowed viewers to select from multiple start times of the same film. Time Warner Cable's chief technologist Jim Chiddix, a former US Army staff sergeant and high-school ham radio enthusiast, was among the category's early champions.

That wasn't all he was thinking about, however. Thanks to connections with Time Warner Cable engineers in Maine and elsewhere, I had gotten to know Chiddix, along with a handful of industry peers who were charged with making smart bets on what was around the corner, technology-wise. From our conversations, I knew they were interested in the budding online computing category.

By late 1992, the leading commercial online service, CompuServe, had topped the one-million subscriber mark, and the top four providers, including the fast-rising American Online, accounted for more than 2.5 million in total. An AOL marketing executive, Jan Brandt, had borrowed a technique she'd implemented in an earlier job with a children's book publisher: doling out product samples for free. Brandt flooded the nation with shiny CD-ROMs, giving away the AOL software and a free month of service that helped AOL blast past rivals to become the world's most prolific online service.

Cable companies, though, were mostly resigned to watching from the sidelines as their longtime antagonists in the telephone industry saw most of the action around online connectivity. Many homeowners began paying for second phone lines to avoid jamming the main line with online connections, giving phone companies a welcome dose of fresh revenue. But what else was there for users to do? There was no other realistic, scaled connectivity option in the residential arena. Cable television's main claim to fame and fortune was still based on television, not data delivery.

That much was apparent from the orientation of Time Warner's Full Service Network (FSN), an ahead-of-its-time interactive media testbed conceived by Chiddix and the executive team at Time Warner Cable. Operating in a suburb of Orlando, Florida, the Full Service Network was a technological marvel, outfitted with ahead-of-their-time computer servers and advanced set-top receivers. Chiddix frequently described the project as a "time machine"—a mechanism for understanding where cable television might evolve in the future. When the FSN formally launched in December 1994, the opening act featured Chiddix and Gerald Levin, then the CEO of the parent Time Warner Inc., pressing a button to summon a movie to a TV screen. The FSN's debut reflected the mood of the moment in cable television: It was still mostly about TV.

* * *

At LANcity, we had brushed up against this same reality for years. In June 1990, angling to make connections in the cable industry, I traveled to its biggest trade show and convention, held that year in Atlanta. On the exhibit floor, Les Borden and I were excited about the chance to show off our technology. We waited patiently behind our rented table in the cheap seats, the rows of smallish vendor displays that were situated on the outskirts of the main exhibit area.

Weeks before, a tireless convention organizer, Barbara York, had sent me a memo expressing intrigue around a new device we were calling a "cable modem." She had never heard of the term and wanted to know more. Although I was grateful we had piqued her interest, most of the trade show activity swirled elsewhere, around elaborate, neon-lit booths showing off glitzy cable network brands like HBO or The Monitor Channel, a new entrant on the scene. Compared with them, we were the little guys.

And yet: Look who was here. One of the senior lieutenants from the mighty TCI. I couldn't exactly make out the name, but a flowing cascade of colored ribbons attached to a neck lanyard made it clear he was a big shot. The TCI executive listened attentively as I told him how our new modem made it possible to whisk data packets down the cable pipe at a breakneck speed—10 megabits per second. I worked in a comment about how we could take the marvels of office LANs and extend them across an entire city. My new acquaintance politely intervened. "Son," he told me in a throaty brogue. "I'm afraid you're at the wrong convention. We do television here. Not computers." And he walked away.

CHAPTER 8

PUPPY LOVE

A key ingredient was still missing: a device that could enable the information highway to become reality.

The dial-up telephone modem traces back to humble roots. Early acoustic couplers that clamped onto telephones—a young Matthew Broderick can be seen hooking one up to a computer in the 1983 MGM movie *War Games*—topped out at 300 baud. Translation: slow

Incrementally, they got faster. But only barely. By 1991, 9,600-baud modems arrived on store shelves, followed by upgrades to 14.4k (or 14,400 kilobits per second) and then the 28.8k and 33.6k modems that prevailed through the mid-1990s. It would take until 1997 for dial-up modems operating over normal residential phone lines to crest the 50 kilobits-per-second mark, with the Chicago company U.S. Robotics introducing a 56k modem as the king of the hill. A U.S. Robotics fact sheet boasted, "With the February 1997 introduction of x2 . . . U.S. Robotics modems became the modems of choice for the Internet revolution."

But the online experience remained agonizingly slow. For users armed with the low-grade modems prevalent in the early 1990s, it would take more than two days to download a standard-resolution feature film—which is why few tried.

This meant that patience was a requisite for going online. The term "worldwide wait" was popularized as a sarcastic stand-in for "World Wide Web" as complaints about the Internet's sluggishness proliferated. "With traditional phone lines, even using the fastest modems now available, many customers find that surfing the Net is as exciting as watching paint dry," commented *New York Times* writer Edmund Andrews in May 1995.

Still, few gave up. There was too much interesting activity happening on the screen. People were becoming captivated by wholly new, often communal experiences. Chat rooms devoted to everything from gardening tips to child-rearing to job opportunities brought together people who otherwise never would have met. Fantasy sports gamers accustomed to waiting for Monday's newspaper could now parse out statistics shortly after the final whistle blew. Correspondence with family members and friends was made commonplace via a new digital transom named "electronic mail." Schools published lunch menus online. Newspaper articles popped onto screens. Celebrity gossip mingled with access to bank accounts and financial transactions. Travel planning intermixed with cooking recipes. Pizzas could be ordered without picking up the phone. People found other people: A 1995 Times Mirror Center survey found nearly one-fourth of online users had an "online buddy" they'd never met in person.

Where videotex providers and their ilk had struggled to gain traction in the US market, the online providers were enjoying a welcome embrace. They were the beneficiaries of fortuitous timing—a convergence of computers, modems, content, and ease of use. Computers, especially, were key. Americans who once wondered what they'd ever do with a home computer were warming to the idea, with impetus inspired by a new crop of computer-based video games and educational software. Online use, however, lagged behind. Only about 5 million adults—this out of a US population of 154 million people over the age of 24—had an online connection at their home by 1994.

Aggravating connection speeds translated to boundaries around what people could realistically do online. On one hand, it was affirming to me to see everyday individuals find something of use in the realm of connected computing. But at the same time, it seemed to me that what people were doing was only a hint at the possibilities. At LANcity, we had bigger ambitions in mind, enabling wholly new accomplishments at home, school, and work: for example, distance learning, where kids in remote school districts could tap into the knowledge of great thinkers and educators; telehealth, uniting patients and physicians over a network connection so people could obtain high-quality care outside of major metropolitan areas; and of course, telework, trading in a tiring physical commute and instead booting up a nearby computer. In all of these instances, slow modems were a gating element.

Sluggish connection speeds and dial-up modems yielded another familiar experience: the temporary transience of "going online." Unlike how data networks performed in an office environment, in the early consumer online realm, there was a starting point and an ending point to an online "session." You came, you clicked, you exited. Online companies simply took this reality for granted—the idea that online sessions were gated, time-defined experiences.

Until it changed its billing practices in 1996, America Online charged users the same way long-distance telephone providers once charged callers: on the basis of time. A per-minute rate codified the understanding that going online meant starting up the payment meter and that disconnecting would stop it. The rhythmic, almost musical intonations of a PC modem during the 1990s became a familiar soundtrack for the decade and later would be preserved as a cherished bit of techno-nostalgia, with modem noises canonized in recordings that remain available for revisiting via numerous websites.

Broadband, though, would be different. Its facility for being "always on" was more than semantics. Here, the flow of digital information was ambient; the online world was less a "place" users visited for a short time than a persistent, at-the-ready presence. "It became part of the fabric of your life, this friction-less resource you could turn to at any moment and have at your fingertips," observed one longtime cable industry technology executive, Jay Rolls. Rolls was among the cable industry's technology vanguard: curious, engaged, and forward-thinking, and not just from an industry posture. As a customer of the regional phone company BellSouth, Rolls had signed up to be a trial user of a LANcity cable modem implementation in the Atlanta area in 1996. It was a revealing moment. "That was my big 'aha,'" Rolls said. "It didn't take but a couple of days to realize, oh my, it's always on." It was the same revelation beginning to occur to everyone.

Rolls, who would go on to work in senior management roles for the cable companies Cox Communications and Charter Communications, was part of a new generation of cable technologists who would rise to prominence at the same time that cable modems began to captivate the industry's collective attention. He and his peers would shortly become our biggest champions.

* * *

Through the 1990s, cable television in the United States historically operated much like a "hive mind," a sort of shared, collective intelligence. The logos on the trucks were different from market to market, the management teams that ran cable companies were distinct, and the investors and families that controlled the voting shares varied. But in an important way, the cable industry tended to move, think, and react to a large degree in near lockstep. City by city, company by company, cable system by cable system, there was far more commonality than divergence within an industry whose participants tended to share with one another an unusual amount of business intelligence—at least compared with other industrial sectors. A well-known media industry investment analyst, Laura Martin of Needham & Co., was fond of mimicking hugs and kissing sounds on stage as she described the workings of a close-knit cable industry.

The all-for-one mentality drew inspiration from cable's Balkanized geography. Unlike warring department store chains, combative automobile manufacturers, or banks vying for deposits within the same downtown business district, the cable industry, for the most part, went about its business without the intrusion of direct competition. The cable company in one town operated without interference from the cable company in the next. It was a friendly patchwork.

The reasons for this unusual business structure reflected brute economics—the intensive capital demands of building a cable system. "It was too darned expensive to have two competitors in a market," pointed out Char Beales, the longtime cable industry association executive who headed the CTAM marketing trade group. "The franchises were basically exclusive."

The very existence of CTAM, a not-for-profit organization with a smart, insider vibe, testified to cable's all-for-one coziness. CTAM staged national press and PR campaigns to take advantage of the shared ambitions of cable industry players that more or less offered the same services and more or less wanted the same things: more customers, more revenue, more attention.

The question now, however, was how to achieve these common objectives. Reflexively pouring more money into more TV channels, a bromide that had worked so brilliantly before, was no longer drawing in as many new customers. Between 1980 and 1990, the US cable industry went from around 16 million connected households to nearly 52 million, for a net gain of 36 million new subscribers and a compound annual growth rate of 12.5 percent. Over the next few years, as the market neared saturation and as satellite TV peeled away customers, growth slowed, with the industry adding 10 million more customers through 1995, a growth rate of only 3.6 percent.

At the same time, there was a new concern rising in Washington, DC. Among the ideas brewing in a possible rewrite of the 1934 US Communications Act was the same concept that would be ordained in a separate 1993 federal court ruling by US district judge T.S. Ellis: that phone companies could use their lines to deliver pay television service, the bread and butter of the cable industry. All told, Congress, the courts, and the FCC were angling to create an unprecedented sort of free-for-all in what had been a very segmented telecommunications market.

TCI's John Malone was quick to act. His October 1993 bid to merge TCI with the regional telephone company Bell Atlantic sent shockwaves through an industry that had long considered telephone companies the mortal enemy. But the tremors didn't last long. Four months after announcing what would have been the largest US corporate merger ever at $33 billion, the deal was off.

There were two culprits. Washington, DC, was one. While Malone and Bell Atlantic's executives negotiated the terms of their merger, the FCC continued to pummel the cable industry with a punitive interpretation of the 1992 Cable Act. On February 25, 1994, one day after the commission announced a new set of rules that would result in further rate rollbacks, the TCI-Bell Atlantic deal officially unraveled. Bell Atlantic's chairman Raymond Smith lamented that "the unsettled regulatory climate made it too difficult for the parties to value the future today."

There was another concern echoing in the background. Besides worries about regulatory conditions, the deal's demise reflected uncertainty over exactly where the brave new world of telecommunications was headed. TCI was supposed to offer Bell Atlantic an on-ramp to a nebulous but intoxicating "information superhighway" that would give the combined company an early lead in bringing the wonders of the digital information revolution. Except, here again, it seemed nobody yet quite understood where or what this so-called information highway actually was. Was it interactive television? Hundreds of channels? Shopping from the living room? Electronic yellow pages? Video on demand? The Internet? All of the above? *New York Times* writer Steve Lohr, in a postmortem about the foiled TCI–Bell Atlantic merger, alluded to unrequited fervor stirred up by technology enthusiasts, including US vice president Albert Gore, who had been talking up the wonders of new technology: "Chalk it up to a dream born in the heady days of last summer—when hype over 'convergence,' 'interactive multimedia' and 'paradigm shifts' had an entire industry spinning fevered visions of the future," Lohr wrote.

Still, *something* was happening. America Online and CompuServe were signing up gobs of subscribers. People by the millions were acquainting themselves with the online experience. Telephone companies had grand designs on transmitting TV channels and exploring new technologies like digital subscriber line (or DSL), which would supercharge ordinary phone lines.

Elsewhere, breakthroughs in the software intelligence that powered wide-area networks commingled with global standards for high-speed network infrastructure being developed via the nonprofit technical professional organization IEEE (which originally stood for Institute of Electrical and Electronics Engineers). But a key ingredient was still missing: a device that could enable the information highway to become reality. A device that could leapfrog the limitations of a slow-motion, dial-up Internet.

That was about to change. LANcity had started making the rounds of cable engineering departments, dropping off complimentary cable modems they could set up and play with for free. We had begun seeding the market, company by company.

* * *

Our version of seeding the market was to let our product do the talking. Not at the executive level of the business, but within the ranks of engineers—people who couldn't resist playing with new gadgets.

We knew just where to find them.

The Society of Cable Television Engineers began life in the 1960s in devotion to the technical work undergirding a multibillion-dollar industry. Three decades later, as we made the rounds with our new LCB modem, the objective remained intact. A telling indicator was the organization's series of technical competitions, culminating at the summertime Cable-Tec Expo, the group's annual trade show and conference. Here, industry technicians—the rank-and-file professionals who make up the lifeblood of the business—competed in a series of proficiency exercises, gaming to be the fastest to splice a fiber line or to rig up a directional signal tap. Winners were written up in trade magazines as if they were celebrities.

The Expo was also the place where chief technology officers and executive vice presidents of engineering gathered on stages to appear at serious-minded panel discussions, such as "Outage Reduction Techniques" and "An In-Depth Examination of Fiber Upgrades." On the convention floor, elaborate display booths festooned with neon signs showed off shiny new line amplifiers and racks of blinking devices. It was nerd-central for the cable television industry. These were our people.

Early on, the data-over-cable idea circulated beneath the radar of the industry's big bosses, the CEOs who ran entire companies. So long as the industry's technologists weren't asking for much money, they were left on their own to play around with new technologies. In effect, these technologists were CEOs of the broadband business—possessing the executive authority to build up the business as if they were entrepreneurs.

We were eager to help, free of charge. Although interest in the possibilities around broadband data was rising, there was still relatively little capital available to experiment with high-speed data implementations among the larger cable companies. Instead, the investment dollars flowing into the cable industry were mostly going to feed the mainstay business of television, along with a few scattered trials of cable-powered telephone services or attempts at (again) restarting a fizzled interactive TV category. If we wanted to influence the hive mind, we'd have to bear a lot of the cost. We'd have to finance the development of our own marketplace.

Thanks to its air of friendly camaraderie, word got around quickly within the SCTE community. "The collegial nature of the industry was a catalyst for this coming about and succeeding," said George Hart, a principal architect for Toronto-based Rogers Communications and a demanding, insistent engineer who would rigorously test our cable modems before green-lighting deployments.

"Collegial" was probably an understatement. The cable television industry worked like a high-school clique in some ways. If you weren't in, you were out. Way out. With its unusual economic structure, cable companies maintained close relationships with one another but often tended to be wary of outsiders. And LANcity was very much the outsider. Who were we? A start-up company from the Boston area, where most of the tech action centered around computers and networking technologies, not cable television. One exception to the rule was a much-admired cable company named Continental Cablevision, headquartered in the Pilot House, a lovely dockside building perched at Boston's Lewis Wharf. Other than Continental, however, the soul of the industry was elsewhere: in Denver, where TCI, Jones Intercable, and the industry patriarch Bill Daniels were based; in Philadelphia, home to a rising power named Comcast; and in St. Louis, where the cable company Charter Communications Inc. would soon take on an interesting new owner, the computer software maven and Microsoft cofounder Paul Allen.

The SCTE circuit was our doorway to these companies and smaller peers. At cocktail-hour gatherings presented by the regional chapters—in places like Rhode Island; Kansas City, Missouri; or Raleigh, North Carolina—someone from LANcity was almost certain to be on hand, armed with a collection of fresh-from-the-factory modems.

Cable's engineering bosses were accustomed to admiring the latest high-frequency amplifiers or diagnostic tools that could hunt down signal leakage. But we were something different: a company most had never heard of before, showing off a cable modem that promised to play a very different role in the industry's tech stack.

Our new friends asked smart questions. Would our technology be able to play nicely with their cable systems? What was the margin of safety in terms of possible interference with TV channels? Which frequencies did we use to send signals back upstream, and how did we deal with noisy upstream environments? How much did our modems cost? And could they try one out?

We explained it all. We talked about real-world deployments in Indianapolis, Amsterdam, Phoenix, and other places where our newly minted second-generation modem was carrying traffic for clients in the government, health-care, and corporate marketplaces. We proudly showed off our new cable modem, miniaturized in a smaller form factor, far more affordable, and thanks to years of development work, built upon proven technology. Most important of all, we gave our modems away—sometimes a pair of our new LCBs and, later, a "six-pack" of our successor third-generation LCP modems, packed into a heavy metal case. We thought of them as puppies; lovable creatures our new SCTE friends could take home and play with.

That's exactly what they did. We handed the puppies off to interested engineers who would take them back to the shop and rig them up, along with our headend translator, to a cable cascade or a makeshift tabletop network. We'd reconvene a few weeks later over the telephone and ask how it all had worked. Almost universally, the reaction was (a) pleasant surprise and (b) a request to keep our modems for a bit longer. Skepticism began to be supplanted by a sense of discovery and delight. Our engineering friends, who had grown up and honed their craft within the traditional "television" side of the cable house, were excited to be playing in a new digital data sandbox, which was fine with me.

I was certain the "Field of Dreams" concept referenced earlier by DEC's John Cyr would prevail. We would build the highway. Over time, entrepreneurs would create amazing digital applications that traveled over it. Our crowd for the moment wasn't cable industry CEOs or business-development types. It was the cadre of chief technology officers and engineers a layer beneath. If we could get their technical blessing and get them excited about our work, we knew the rest would follow. We were educating and empowering an industry from the ground up.

It was a small circle. Around the time our LCB modem was introduced in the spring of 1993, there were relatively few engineers or executives across the entire cable industry who were deep into exploring the possibilities of high-speed Internet delivery. They were people like Michael Giobbi, the chief technology officer from the Pennsylvania cable company Armstrong Communications, who seemed to possess an early sixth sense about where the world was going and how the cable modem could get it here. Ditto for Shawn Fisher, an operations executive for a midsized cable company, Bresnan Communications. After receiving an invitation from Fisher, I showed up at the company's White Plains, New York, headquarters, where I was surprised to see the company's founder and CEO, Bill Bresnan, and his brother, Daniel, seated at a boardroom table. I'd expected only the engineering team; this was an exception to the rule. Both men were keenly interested in this new concept of high-speed data running over cable systems and, on the spot, ordered up a six-pack of modems to run through the paces. This was an unusual instance. Cox Communications and its CEO Jim Robbins, a mentor of mine, made for another unique story—where a top executive seemed to be ahead of the game.

The more people we met, the more I understood how the industry was structured and where the power was. The cable business was bifurcated. On the one hand were hundreds of smallish, independently owned cable companies—companies like Cable Alabama, an early adopter of our technology, or Nebraska operator interTECH and its president and CEO Bill Bauer. These independent cable companies and their like-minded peers represented the old guard of the industry, often individuals who had financed their cable TV ambitions by taking out mortgages on homes, borrowing money from family members, and in one case—that of the mercurial founder of Jones Intercable, Glenn Jones—hawking an aging Volkswagen. But the pioneers wouldn't persist forever. Beginning in the early 1990s, a wave of acquisitions and asset swaps had begun, with many smaller and mid-sized companies exchanging territories or selling their systems to a bigger consolidator—industry titans like Malone's TCI, the Boston-based Continental, or the fast-growing Comcast Corp.

Among the larger survivors was Long Island–based Cablevision Systems, a cable provider founded by the industry legend Charles ("Chuck," to friends) Dolan. Dolan was a quiet, slight-framed Bostonian who carried tremendous weight in the industry. He had a hand in plenty of seminal moments, including the origins of the premium TV service HBO and the creation of popular entertainment channels like American Movie Classics. Dolan's top technology executive was Wilton Hildebrand, a salty-tongued fan of LANcity and one of our biggest champions—except, not so much as a paying customer. Hildebrand would frequently ask me for demonstration models of our modems but steadfastly refused to commit to purchase orders. Still, he became part of our trusted sounding board, an unofficial company advisor and an influential mentor of mine.

I met Hildebrand, "Wilt" to almost everyone, for the first time at one of the big annual SCTE conventions, where a crowd of admirers—mostly sales representatives for prominent equipment makers—gathered around him. Anybody who sold cable television equipment for a living—the well-dressed, polished representatives for big companies like General Instrument, Scientific-Atlanta, and Texscan Corp.—knew where purchasing orders would originate. Armed with American Express cards, they competed for the affection and attention of decision-makers like Hildebrand.

In those circles, we were all but invisible. Our approach wasn't to wine-and-dine the industry's CTOs. For one thing, I couldn't afford it. And second, our technology told a better story than any brochure or fancy dinner could convey. We'd leave all that to the big guys with the big Amex tabs. Instead, we worked the marketplace by volunteering to help people begin to understand where this high-speed data revolution might be going.

Shortly after we'd started manufacturing our second-generation LCB modems, Hildebrand had called me, demanding—politely, but still demanding—that I show up the next day in Long Island. He wanted to show off the possibilities of high-speed data connectivity over cable to Cablevision's senior executives, Charles Dolan among them.

I was happy to comply and grateful to Hildebrand and his colleagues, not just for paying attention to the work our small company had been doing but also for playing a behind-the-scenes role in our development work. Among the cable systems that Cablevision Systems had acquired was a familiar name: Nashoba Valley Cable, the small-town cable operation in Stow that had provided us with an important testbed for our modem technology. At Hildebrand's behest, I arranged to fly a small team to Long Island the next morning, armed with a case full of puppies, ready to help and confident we were aligned strategically with a personal computer revolution we were certain would win the battle for interactive media supremacy.

CHAPTER 9

SCREEN WARS

Somehow, the TV set had to either become a computer or be attached to one.

One of the signature techno-gadgets of the mid-1990s was the Tamagotchi: a hand-held swirl of plastic encasing a coarsely pixelated, monochromatic screen on which kids tended to the needs of a digital pet hatched from a digital egg. Attentiveness to the creature's daily needs was a requisite. The well-cared-for Tamagotchi cooed and chirped with happiness. The neglected Tamagotchi died and was succeeded by a new hatchling—along with a pledge by the owner to do better this time.

The Tamagotchi (its name a mash-up of Japanese words for "egg" and "watch") was a telling artifact of a period when the world was on the cusp of tremendous technological breakthrough. Although computerized toys were still relatively crude, it would take just 10 years (from 1997 to 2007) to leap from the peak of Tamogotchi popularity to the iPhone. And in the background was the device that influenced the whole of the consumer tech revolution: the home computer. In 1989, only around 15 percent of American households had a computer, with early adopters mostly using the machines to send emails or to play games. Four years later, it was 23 percent; by 1997, 36 percent; and by the end of the decade, at a tipping point: More than half of US homes were computer-enabled.

Rising computer penetration was a key propellant for the adoption rate of online services, which had been limited by unimpressive computer graphics capabilities and slow processing speeds. By the early 1990s, computer processors like Intel's Pentium had improved markedly, paired with more computer memory to allow machines to accomplish more, quicker. Another contributor was online readiness. By the mid-1990s, popular home computers were coming off the assembly line with a built-in phone jack (and later, a network card). Manufacturers understood the appetite for connecting to an emerging online world.

Even so, the computer had to overcome a challenge for in-home screen supremacy throughout the 1990s: Almost everybody had a TV set. Thus, a battle. On one side: architects of an information revolution who believed the home computer would be the instrument of a new ecosystem. On the other: believers who clung to "interactive television" as the path to glory.

Faith in interactive television stubbornly persisted despite dozens of false starts. Starting with a gimmick associated with the 1950s kids' TV show *Winky Dink and You*—the idea was to place a transparent film onto the TV set screen and let kids draw bridges and roads the character would step across— developers had tried again and again to turn the TV set into an instrument of active viewer response, but to little reward. Interactive television, per the *Chicago Tribune* writer Mike Langberg, was "the punch line of a bad joke: It's next year's technology, and always will be."

Still, dreams die hard. WorldGate, a mid-1990s creation founded by a well-known cable industry technology executive, Hal Krisbergh, sought to forklift the Internet from the personal computer and deliver it to the TV screen. The same construct was behind WebTV, a dial-up, TV-meets-Internet service that began life in space rented from a Palo Alto, California, car dealership and was later acquired by the tech giant Microsoft.

However, there was a core problem with reviving the interactive TV dream. Enlivening the TV set with fanciful graphics and quick-response interactivity would require a powerful computing system. Somehow, the TV set had to either become a computer or be attached to one.

That was problematic because computers—at least powerful computers— were expensive. Retail prices for personal computers advertised in 1994 ran from around $1,200 (Compaq ProLinea 425) to more than $2,400 for a new Toshiba notebook machine, well beyond the per-unit costs cable companies could tolerate at scale. Even with generous volume discounts, outfitting the millions of customers served by cable giants like TCI or Continental Cablevision with off-the-shelf PCs would have been ruinous. Because of this reality, the interactive TV dream was compromised by economics: a dependence on low-grade, inexpensive digital cable boxes that were multiple generations behind the prevailing PC standard.

To wit, the entry-level digital cable box manufactured by General Instrument in the early days of cable's digital video transformation featured roughly the same amount of random access memory (or RAM) as what Apple had injected into its debut computer, the Apple II, eight years before. As for processing muscle, General Instrument's DCT-1000 box, introduced in 1993, used a 32-bit Motorola chip with a clock speed of 16 MHz—the same specifications found on a chipset Motorola brought to life in 1987. Translation: It was way behind the times.

It's not that the cable industry and its technology suppliers lacked savvy. General Instrument and its archrival Scientific-Atlanta Inc. were longstanding technology companies with large development staffs, global sourcing experience, and established manufacturing relationships. They were entirely capable of building high-end cable boxes that rivaled the processing power, speed, and memory of a modern PC. If, that is, cost wasn't a consideration. But it was. In order to make the cable industry's digital television transformation work at scale, the first-generation digital cable boxes had to be priced at less than $400 apiece—a big drop from the $3,000 per box Time Warner Cable spent on its Full Service Network.

A few companies were game to try. Microsoft Corp., bent on making interactive TV a reality, worked feverishly to develop a more powerful box that could realize the WebTV dream, blending familiar television channel surfing with new online capabilities. A similar play was in the works at Oracle, the software giant, whose affiliated company NCI was putting together an ambitious plan to swoop into the cable set-top box category with a souped-up digital box. Behind these gambits was a faith that new-age set-top boxes, not home computers, would emerge as centerpieces of a new connected lifestyle.

That much was apparent in February 1993 when *The New Yorker* published an 8,000-word article in which two media industry mavens, the Hollywood impresario Barry Diller and the cable industry chieftain Brian Roberts of Comcast, rhapsodized about the marvels of newly acquired Apple PowerBook computers. The article by the dean of media business writers, Ken Auletta, was revealing: Although both men had become enchanted by the things they could accomplish over their new laptops, both seemed to believe the television set—not the computer—would become the instrument of choice for fulfilling the connected highway dream. Auletta quoted Diller, talking about "a powerful microprocessor in a cable-converter box inside or beside the (TV) screen."

There it was again: TV sets. Channels. Old habits die hard. Even as the personal computer revolution gained steam, media moguls found it hard to give up the interactive TV dream. That wasn't exactly good news for LANcity. Our thesis was to take advantage of the steady march of the computer as the primary mechanism for interaction in the home. I recognized early on that interactive TV platforms would pale in terms of functionality, availability of

Internet-based services, and—important point here—an ecosystem that lived atop open standards, inviting innovation and scale.

It was impossible for me to fathom how interactive TV sets manipulated by push-button remote controls were going to usher in the age of distance learning, telecommuting, or digital healthcare. Besides that, the screen resolution and the graphic processing prowess of a typical TV set were inferior to the display of modern computer monitors. "There was just this physical aspect that became the barrier for interactive TV to become anything more than a plaything," observed my colleague George Hart from Rogers Cable. Hart and others believed most people seemed to prefer to use television as an instrument of relaxation and entertainment—not a vehicle for emailing bosses, conversing in chat rooms, or tracking stock prices. Television was innately a passive "lean-back" technology. Computers, in contrast, were much better optimized for a "lean-in" experience.

Happily for us, consumers proved to be just as reticent as I was. Why pay several hundred dollars for a device like Microsoft's WebTV box, people wondered, when a more powerful, capable device—a home computer—was already present and ready? Here again, the ghost of the videotex era reared up. Try as they did, the new interactive TV aspirants could not overcome familiar problems, including the absence of open standards and flexibility in developing software. The same sorts of application enhancements developers could produce in a few months in the home computer ecosystem could take two or more years in the hazy interactive TV environment. Open standards were infusing the world of computing with a critical scale that invited rapid innovation as developers rallied around common approaches for innovating. No such harmony prevailed in the siloed world of interactive TV.

There was also the issue of cost. "If money is no object, we can make interactive television work," a Comcast senior executive, Mark Coblitz, told the industry periodical *Multichannel News*. Coblitz, a keen-minded strategist who would become instrumental in convincing Comcast to invest heavily in broadband data delivery, was a realist. "But money is an object, and I don't think we know how to do it in a way that scales across hundreds of thousands of customers."

Coblitz and others recognized that besides being more powerful, computers were rising to become everyday appliances with presence and scale across a majority of US homes. In contrast, by 1997 the percentage of US homes that were outfitted with one of the new breed of cable digital set-top boxes lagged far behind, in the low single digits. "The interactive future that Malone had so cockily predicted back in December 1992 was inching further from his reach," wrote the former *Wall Street Journal* reporter Mark Robichaux in his Malone biography. "All around the country, cable giants grounded their interactive TV tests citing delays, high costs, and concerns about whether consumers would really pay more money for fancy new services."

Still, the back-and-forth battle between adherents of interactive TV and those who believed in the computer had a hidden benefit: The momentary uncertainty gave LANcity a welcome moment of grace in which to develop and optimize our own platform. Had interactive television taken off sooner, it's possible our work could have been short-circuited.

In the background, some cable industry technology experts had recognized the importance of the computer all along. TCI's longtime technology executive Tom Elliott had been advising his cable industry colleagues for years to take the computer revolution seriously. "Guys, this is coming," Elliott had warned peers when he was championing the ahead-of-its-time X*Press computer service in the 1980s. "These PCs are not going to go away . . . They are going to find themselves in all your customers' homes."

Elliott was prescient. TVs were about leaning back. Computers were about leaning in. And if that realization wasn't enough to propel the home computer to a place of superiority, a new acronym beginning to take hold just might: www.

* * *

A few blocks from the eastern edge of the Harvard University campus, near the intersection of Broadway and Ellery St., the Cambridge Public Library beckons visitors with a checkerboard walkway of maroon and gray bricks, leading to the building's glass door entryway. Inside, a wireless data network volleys electronic signals to computers and digital devices at multi-megabit-per-second data rates. No surprise there: Connecting library patrons to the World Wide Web and to a trove of information networks is a capability visitors routinely take for granted. It's no less expected than electricity—or books on shelves.

At least, that's the case today. But it wasn't the case in the summer of 1993, when Harvard University graduate David Fellows was putting the finishing touches on a novel technology implementation at this same library. Fellows was angling to make the Cambridge library one of the first locations in the United States to connect users to the Internet at a speed few people had ever experienced.

Fellows was a tall, energetic man, polite but intensely determined. In 1976, he had captained the US men's rowing team in the Olympic Games. He had joined the Boston-based cable company Continental Cablevision in 1992.

Recruited by the company's president, Bill Schleyer, Fellows had impressed the company's cofounder, Amos Hostetter, enough to be named the company's first corporate engineering executive. Hostetter and his right-hand man, Schleyer, wanted a front-row view of new possibilities for transforming their company's technology path. They found their candidate in Fellows, a former GTE research scientist who had been running the transmissions systems unit of the cable equipment maker Scientific-Atlanta.

Like most of his peers on the technology side of the cable industry, Fellows had been steeped in the traditional video side of the business, manufacturing and selling equipment to enable the delivery of television signals. Here with Continental, however, was an entirely different proposition that excited Fellows: fiber optics. To Fellows, the idea of lacing cable systems with fiber seemed hugely promising. Exactly what applications and functionality would rise to the top wasn't known. However, it was clear that fiber could usher in a new era of interactive communications and other services.

I was very familiar with Fellows by early 1994. We'd met at least a dozen times, visiting either at Continental's historic Pilot House building or LANcity's office. Continental had already been using our first-generation modem to connect a handful of corporate and government clients in Manchester, New Hampshire, including the tech entrepreneur Ross Perot's Electronic Data Systems. Continental later would follow up with one of the earliest trials of high-speed data over a residential cable network. The more we talked, the more I sensed that Fellows shared my vision of an impactful opportunity, one that could conceivably grow to involve tens of millions of users.

The foiled dreams of interactive television had left a deep impression on my new colleague. Tracing to his days at GTE, Fellows had gotten to know the developers of a GTE-backed interactive television service branded "mainStreet." This successor to the videotex experiments of old relied on leased 6 MHz cable channels to offer a grab bag of interactive TV capabilities, like summoning restaurant menus to the TV set. Fellows was never a big mainStreet believer; he was skeptical the service could find a way to scale up to enable massive usage. But conversations with the developers behind mainStreet, a Cambridge company called Delphi Data Services (later gobbled up by Rupert Murdoch's News Corp.), had led him down a different path. In private conversations, the Delphi team had convinced Fellows that a computer, not the TV set, would ultimately rule the category.

The timing here was serendipitous. Around the same time he huddled with the mainStreet team, Fellows was captivated by a new Internet application called the World Wide Web and had been considering using Continental's cable lines and our LANcity modems to connect users to it. Until now, the prevailing model had been to connect a computer user to an online exchange either through a dial-up commercial online service like CompuServe or through the information technology department of a nearby university. The university would then connect the user to the broader Internet using a tool like Gopher, an early interface for finding and retrieving online documents.

Not long before, Fellows had sat in on a seminar presented by Harvard University's Kennedy School, where he became entranced by seeing a beta version of the iconic Mosaic web browser. The concept of "hyperlinking"—allowing users to sprint from one page to another across multiple connected servers—was fascinating. "I took one look at it," Fellows said, "and I thought: My mother could do that."

A similar recognition was occurring elsewhere, giving credence to the idea that with high-speed data service, the cable industry might be onto something important. In Toronto, where the cable company Rogers Cable had begun testing cable modem technology in its engineering laboratory, the web browser loomed large. George Hart, the principal architect with Rogers, was similarly smitten. "You'd open a document and there would be a hyperlink to something else," he marveled. "You could spend hours exploring."

Continental Cablevision had been early to the data-over-cable game, connecting a handful of institutional users over cable lines starting in 1989. Executives including a New England–region vice president, Kevin Casey, along with Schleyer, now were reasoning that the same technology might be used to connect individuals, not just organizations, to the Internet and to the emerging World Wide Web.

The public library looked like an ideal launch point. Not only was there a natural affinity for research, but Fellows, conscious of the need to seize on a profitable business model over time, figured a library could act as a sort of evangelistic force for this new broadband cable technology. The idea was that patrons seduced by high-speed Internet access at the library might want the same capability at home. In concert with an Internet service provider named Performance Systems International, Fellows and a quartet of brainy technologists from the Massachusetts Institute of Technology rigged up one of the first public cable modem implementations in the world.

The library was connected. All it took to give it a whirl was a seat at a computer and a library card. We had passed an important test. Satisfied with our technology's performance and appreciative of our pricing philosophy—we were passing along anticipated volume manufacturing breaks before we actually had them—the Continental team followed up with residential broadband implementations in Newton, Massachusetts, and in a few neighborhoods in Cambridge. In fact, somewhere along a street in Cambridge is one of the two or three longest-running cable broadband residential addresses in the world. It was connected in 1993 and, as far as I know, has stayed online ever since.

Then, the biggest project of all, the one that put cable modems on the map nationally, at nearby Boston College. In 1994, Continental's engineering team began working with the college's IT department to connect residence halls, classrooms, and offices to an ahead-of-its-time cable modem implementation. The numbers were astonishing for the time. At its peak, the project connected more than 6,500 dorm rooms to the Internet at high speed, thanks to LANcity second-generation cable modems tucked inside hallway electronics closets and connected to Ethernet cables.

Boston College quickly became LANcity's largest network, surpassing even our Rock Island Arsenal implementation and its 5,000 users. By the fall of 1995, a college that had been founded by Jesuits 130 years earlier became the world's most prolific deployment of high-speed data, running over a hybrid fiber-and-coaxial cable network. The Boston College deployment was larger even than many of the nation's small-town cable systems. A few thousand undergraduates, in between attending lectures and scratching out late-night term papers, were experiencing the cutting edge of broadband data technology, enjoying multi-megabit per second data rates that had until now been confined to business offices and their local area networks. Of note, the implementation coincided with the birth of music delivery over the Internet, with file-sharing standards like MIME enabling fans to download and swap songs with new-found ease.

In Boston College, Fellows and Continental had discovered an elegant simplicity: One customer to bill, an established wiring system in place, and an existing acceptable use policy that spelled out clearly the university's terms of Internet access. The national press took note. The *Wall Street Journal* reported that usage results from the deployment Continental had nicknamed Project Agora (a nod to the ancient Greek public gathering place) suggested "high-speed access increases the likelihood that users will become addicted to surfing the Net."

The following year, not long after Fellows and his team of MIT graduates established one of the cable industry's first large-scale backbone data networks—shuttling speedy data streams from Florida to New England—Hostetter would agree to sell Continental to the Denver-based phone company US West. Once the ownership transition was complete, Fellows found himself facing off with a phalanx of US West executives who peppered him with questions about the emerging high-speed cable broadband opportunity. Exactly how many users could the network sustain at one time? What were the peak traffic rates? How many people had PCs? What applications did they use? How many of those people were already online, using a commercial service like AOL? Fellows greeted every question with an open admission: He didn't exactly know. Not that Fellows and his team were naive. They had kept a close watch on the macro-level metrics, like bandwidth consumption over the network. However, precise details on consumer behavior were to be determined.

The explanation flustered the US West team. Why in the world, an exasperated executive asked, did Fellows launch the service without first divining these essential answers? His answer was drawn straight from the cable cowboy script: "I figured I wasn't going to get any smarter," Fellows said, "by *not* launching it."

CHAPTER 10

ROSE GARDENING

Our baby was born. The world was about to change.

In the autumn of 1989, we had proven that with some careful adjustments, we could move digital data over a real-world residential cable system without bothering the adjacent TV channels. Five years later, we were back in the town of Stow, Massachusetts, connecting our second-generation modem over even longer distances.

I had no reason to suspect we'd encounter trouble. Our QPSK modulation approach, a complicated brew of algorithms we had been working on since 1988, had proven capable of burrowing through noisy upstream pathways. Our newly patented Unilink-II protocol had been simulated to an exacting degree in our lab, where we'd devised a replica of a cable system. It also worked flawlessly in field implementations involving numerous private cable systems. And we'd already proven a previous iteration of our technology worked just fine in this same town, connecting two residences over a two-mile run of cable.

Back at the office, we were close to greenlighting a new integrated circuit for our forthcoming crown jewel, a residential cable modem designed for mass users over long distances, ready to rock the world. A blueprint for our new chip was scheduled to be fabricated—turned into a physical chipset—in just 10 days.

I was sitting at my desk when my mobile phone chirped. Nick Signore's name flashed on the screen. My longtime colleague sounded worried.

"Uncle Rouz," he told me in a sober voice, using the nickname he'd invented for me a few years before. "Something's not right."

* * *

The phone call was a body blow, utterly confounding. I knew Nick wouldn't bother me at this point if things weren't serious. But in Stow, he explained, he was picking up troubling signs. Nick had been dutifully taking readings at connection points after every amplifier. The farther he traveled, tracing the path of coaxial cables strung across utility poles, the worse the signal became. Until it became no signal at all. "It just isn't here," Nick told me.

I was stunned. Our second-generation modem had performed flawlessly not just in the lab but in numerous field deployments. Just ask the students living in Boston College dorm rooms. The only difference here was distance and the fact that we were operating over a residential cable system deep into a neighborhood.

Welcome to the phenomenon called "tilt."

It's shorthand for a sort of dancing act that involves amplifiers and the frequencies into which they inject power. Everybody in telecom knew that signals tended to attenuate more at higher frequencies than at lower frequencies; it was simply a rule of life. That meant the signal associated with a lower-frequency channel requires less amplification than a higher-end channel—even when the two channels are traveling over the exact same length of cables. Imagine plotting a graph where the amount of signal decreases or "tilts" downward as the channels climb higher in frequency along a cascade of amplifiers. The remedy was a logical one: add specialized filters inside amplifiers that would compensate for the frequency slope, minimizing any impact on signal fidelity.

For video channels, filters neatly solved the tilt problem. But as we were now discovering, these same filters produced slight variations, amplifier to amplifier, involving signal delay. Until now, nobody much cared. For cable TV channels, these "group delay variations" barely registered at all. The typical amount of signal delay was no more than 21 nanoseconds per amplifier—21 billionths of a second. So by itself, an individual amplifier caused almost no meaningful consternation. Movies on HBO passed through the amplifiers and cable lines and onto a customer's TV set just fine.

But for data streams, group delay variations could be vexing. Over the long amplifier cascade Nick was working with, so much timing variation had been introduced that the signal was badly distorted. As Nick had observed, over longer distances, the data stream that was supposed to be the grand achievement from my 1987 vision had simply . . . vanished. It was gone.

Reengineering cable systems with long fiber-optic runs to replace lots of amplifiers was only in the early innings in 1994. Ripping out coaxial cable lines and replacing them with fiber optics on a large scale would take decades.

Instead, lengthy cascades of multiple amplifiers remained the rule. Once again, the town of Stow had presented a microcosm of what we were certain to encounter across the broader cable horizon. We'd have to find a way out of the mess we were now confronting. And fast.

Nick's phone call came on a Friday afternoon. Our new chipset was due to hit the fabrication line the following Tuesday. I called my contact at the AT&T fabrication facility, explaining what had happened and asking—pleading, really—to swap our production slot with another customer.

Our timing wasn't ideal. A computing revolution was under way across the nation, causing demand for integrated circuits to become so intense that fabricators were running around-the-clock schedules, piling on projects one after another, mass-manufacturing chips the way fast-food restaurants churned out hamburgers. Thankfully, I had a sympathetic companion. My contact at the AT&T chipmaking facility had become intrigued by what we were doing and, by now, was cheering for our success. The more I shared about our objective, the more he seemed to sense we were onto something important. But the schedule was already overloaded. The best he could do was delay our deadline by one business week.

I gathered the team, caucusing with Bill Corley, John Ulm, Paul Nikolich, and Chris Grobicki. All four agreed to come into the office early on Saturday morning. I was relieved when our chip-master Kurt Baty answered his phone. He wouldn't be able to make his way to our office until Sunday. But Sunday was better than Monday. We agreed to work on the problem in a frenetic burst of activity.

I had to alert our contacts at DEC, as well. Nervous about disappointing an essential ally, I huddled in our office—we'd established a "situation room" to deal with the problem—with John Cyr. I explained our dilemma. John understood. He was a technologist with a technologist's viewpoint, respectful of the reality of our situation. He agreed to an extension that would have otherwise violated the terms of our agreement. Cyr trusted our skills and our application of logic. He recognized there was no point in manufacturing a chipset that would imprint a compromised protocol into silicon.

It was Bill Corley who came up with a novel idea. The thought was to correct for the latency phenomenon we were experiencing by compensating for the effect elsewhere. At the center of his remedy was a "least means square" algorithm that evaluated the difference between a desired signal and the actual signal. The concept was to determine exactly how much group delay distortion the cable system was injecting. Once we knew, we would intentionally pre-distort our signals with an advance group delay error—the inverse of what the amplifier filters could be expected to cause. This processing technique, known as equalization, was meant to pull off some sleight-of-hand trickery, effectively canceling out the delay errors. Innocent-seeming filters installed in cable television amplifiers got us into this mess. Mathematics might get us out.

We worked into the early morning hours to integrate the newly conceived algorithms into our code. To try to identify any potential failings in advance, we flushed torrents of digital data through our cable modem, blasting packets from 30 workstations we'd synchronized to emulate the sort of traffic we would encounter in a real deployment. We ran the workstations for days. Our tests—data in, data out—lasted only about 30 seconds. But they touched and toggled close to one million logic gates in our chipset. We found problems, fixed them, and reran the test. Over and over. Until we ran out of time.

On the afternoon before our deadline, Kurt Baty applied the final elements of our new code to a pattern generation tape saved on a computer hard drive. Nothing was certain. I knew there was no way we could have accounted for every possible permutation. We couldn't be sure we'd corrected the group delay problems until we had our new chips in hand. Either they would work, or we'd blow past a critical window that was vital to LANcity's future.

Privately, I figured we had about a 30 percent chance to come out okay. But 30 percent was better than nothing. I raced to Boston's Logan International Airport that evening, barely making it in time for the last Federal Express flight, which took off nightly at 11:30 p.m. I drove to deliver the package myself because I knew better than anyone the quickest, most direct route to the Federal Express office. I handed over a cardboard box containing our hard drive.

Once again, we waited. Once again, our future was at stake. Seven years and counting into our journey, we found ourselves in a familiar place: at the edge. But this was nothing new. There had been countless challenges sketched out in ink from blue markers across hundreds of sheets of oversized paper. Every time, we applied combinations of ingenuity, imagination, and physics to see our way to the other side.

I hoped some variation of this same blend would see us through now. Already, I'd been turning over in my mind the talk I'd need to have with John Cyr and my colleagues at DEC in the event that the new digital filtering approach failed. I tried to calm myself. Not everything was gloomy. Even if we had to go back to the drawing board, I still believed we were ahead of the game, with a 12- to 18-month lead over what anybody else might be able to accomplish.

I thought back about how far we'd come. We had been connecting data users over coaxial cable networks since 1988, getting an up-close-and-personal look at the vagaries and peculiarities of the cable television infrastructure as they related to data traffic. Nobody else had their version of our Rock Island Arsenal experience and the lessons we learned. Nobody else had undertaken the important step of roaming the streets of Stow, Massachusetts, working in a real-world residential cable environment to stress-test their technology. Our competitors were working in laboratory settings. They weren't scurrying up utility poles, ascending in bucket trucks, and unscrewing rusted bolts on amplifiers. They hadn't gathered field data and used it to make corrections to their cable modem design. They hadn't discovered oddities like we had encountered not long ago in Long Island, New York, where Cablevision Systems had been puzzled about why the high-speed data network was failing almost every evening starting around 5 p.m. near a particular intersection of busy streets. Our investigation found the culprit: Truck drivers idling at the stoplight were communicating over Citizens Band radios that emitted high-power signals into a low-frequency band, causing interference with the upstream data network. Poorly tightened physical connections on a nearby length of cable allowed the radio signals to seep into the same frequencies that conveyed Internet data—another example of the perils of ingress. After tightening up the connections, the problem was solved.

And now, so was our latest dilemma. Working within a test bed we'd created—we called it RDA: our rapid, detailed, automated environment—the tests of our newest chip demonstrated my 30 percent calculation had been pessimistic. The new chip, encompassing Bill Corley's equalization magic, proved to be fully functional. We hurled massive amounts of data through our new modem, combining dozens of workstations to produce a sort of digital blast furnace capable of slinging data across 200 modems running over the coaxial cable system in our lab. Back in Stow, we returned to the scene of the crime to make sure the technical and distance issues that had bedeviled us were subdued. This time, success: Our data streams were performing exactly as intended, hurtling data streams without incident across miles of amplified cable lines. The innovation Bill devised would later be described in US Patent 5,608,728: a method for equalizing the forward and reverse channels of a communications network. It would soon become a cornerstone remedy for making high-speed data transmission over residential cable reliable, affordable, and practical.

Thank heavens. Thanks to mathematics and physics. Thanks to everything and everyone: LANcity's original team and the new employees, advisors, customers, and vendors who had collectively made broadband over cable real. The broadband dream was still alive. Our baby was born. The world was about to change.

* * *

Around the time we found a remedy to our tilt issue, the cable industry was enjoying a moment of reprieve on the regulatory front. Reed Hundt, the FCC chairman who had bedeviled the cable industry by imposing severe rate restrictions, had begun to see the cable television business in a new light. Hundt was starting to realize the 1992 cable re-regulation law, celebrated by Congress and the commission as one of the great consumer protection acts of all time, hadn't done very much to ease the day-to-day economic burden on consumers. "Most Americans, indeed, never knew that their cable bill had been lowered by, say, $1 a month," Hundt pointed out. Moreover, whatever savings had accrued often were plowed back into the cable company as consumers bought an extra premium service or two. So the net effect was hardly the everyman financial benefit that US Congress had hoped for. This change of perspective, coupled with continuing lobbying from cable's association representatives and company executives, provoked an important new change of heart: Hundt agreed to implement new "going forward" rules that would provide some daylight for the cable industry.

The bargain went like this: Cable companies would still live under the burden of strict rate regulation for their lower-level channel tiers but could enjoy newfound freedom at the higher end of the menu. They could arrange for new add-on packages that would cobble together many of the industry's more popular channels and present them without government rate oversight. For these new "tiers" of channels, cable companies could charge whatever they wanted—and whatever seemed likely to attract the most customers.

The freedom to add new packages and charge extra for them sent relief coursing through the industry. Almost immediately, creative new packaging approaches came into vogue, as cable companies and their programming partners realized the new freedom would create a gateway to growth. An industry that had been badly crimped by federal rate oversight had new life.

New life was exactly what the industry needed in order to undertake new technology in the United States. But internationally, some of the more interesting action was already happening. Cable companies like Amsterdam 2000, London-based Telewest Communications, and others were among LANcity's biggest champions, motivated by a sense that cable was on the verge of a transformation. For a London technologist named Graham Sargood, that epiphany came from watching schoolchildren in a London classroom.

Sargood had joined Telewest in 1994 after a stint with AT&T Network Services. Hired to explore the broadband-over-cable opportunity, Sargood got busy fast. Within a few months of his arrival, Sargood's engineers were putting the finishing touches on LANcity cable modem implementations for a quartet of primary schools near the Scottish capital, Edinburgh, and in southeast London. The idea was to tease out findings about how students might engage with an online connectivity service that was worlds apart from what anybody had ever known.

Sargood, angling for a close-up view, arranged to visit one of the London schools, where the headmaster was exploring ways to take advantage of the Internet's distance-learning promise. Stationed alongside a wall in the school's computer lab as students filed in, Sargood realized almost instantly that Telewest had a hit on its hands. The students were transfixed as they whisked from one online destination to the next, pages materializing instantly, "They were so blown away," he remembers. "That was really when the penny dropped, and you felt: This is going to be big."

Not that everything came easy. In England, Sargood's engineering associates were about to climb a steep learning curve. For one thing, customers often powered down their cable modems at night out of habit, requiring the network to go through an elaborate process of re-registering their IP addresses when they returned to life the next day. Installations, too, could be tricky. Without the ubiquitous Wi-Fi we take for granted today, computers had to be tethered directly to connecting cables, limiting the physical locations where modems could be planted inside a London flat or a country home. There were also internal issues: The back-office software that supported customer relationship management and accounting for the cable company wasn't designed with a new Internet service in mind. Telewest worked around that obstacle by setting up its Internet offerings as a separate business unit, with its own back-office systems. The name they adopted was as straightforward as could be: Cable Internet.

Workarounds or not, the Telewest deployment marked an important inflection point. We'd been doing business on a smaller scale with other Western European cable companies, but here was a high-profile player, jointly owned by John Malone's TCI and the phone company US West, that was using LANcity modems and our headend translator to power an entirely new business line. Sargood had put his own professional reputation on the line by allying with a little-known tech company from Massachusetts. But he'd done his homework. He was certain that our modem technology was at the head of the class. And that broadband-over-cable was going to explode. Sargood became a willing evangelist for the category, appearing at conferences and in front of peers to share details about Telewest's deployment and experiences with wholly new technology.

* * *

At this point, the big US cable companies were hopping aboard the same train. The industry analysis firm Paul Kagan Associates reported that at least seven US cable companies were testing cable modems by 1994. Among them was my friend Wilt Hildebrand's Cablevision Systems, which had trials underway in Yonkers and Long Island, New York. Our Boston neighbor Continental Cablevision was using our modems in Cambridge. Cox Communications, the big Atlanta-based cable company, had identified Orange County, California, as the entry market for its modem experimentation, also giving the nod to LANcity for the technology. And in Tampa, Florida, the Denver-based cable company Jones Intercable had started to hook up our cable modems to its hybrid fiber-coaxial network. Not long before, I met with the top technology executive for Jones Intercable, Chris Bowick, at the company's headquarters. Bowick casually informed me that Jones would be using LANcity's modems for various trials, including what would come to be a hugely influential deployment in Alexandria, Virginia. I was elated at the news.

Another important entrant was the cable company Viacom, which had started connecting a few "friendly" households (most belonging to employees) in the Silicon Valley community of Castro Valley. The installation site had been chosen as Viacom's answer to Time Warner Cable's Full Service Network—a modernization that would allow for new interactive television services, a major channel lineup expansion, and the transmission of Internet traffic at high speeds over an available cable channel.

There was also another important player in the mix: Times Mirror Cable Television. In Phoenix, Arizona, Times Mirror was deep into the most sweeping cable modem implementation yet, working with LANcity, our partner DEC, and Arizona State University to connect several large aerospace companies to a fast, cable-powered data network. Inaugurated in late 1991 as a smallish project called ValleyNet, the project grew fast, emerging by late 1993 as one of the first big commercial implementations of a long-distance data network running over cable lines and serving multiple points of presence.

By the summer of 1993, Times Mirror had renamed the project as the Electric Commerce Network (EC-Net for short). The deployment linked power users like McDonnell Douglas and Lockheed to Times Mirror's newly upgraded hybrid fiber-coaxial lines, with the longest run extending for close to 15 miles.

The companies were collaborating with one another across space using applications like computer-aided design, video conferencing, and electronic whiteboarding. Designers from McDonnell Douglas's helicopter design group sent computer-aided drawings across the network while communicating through an emerging application called electronic mail.

Martin ("Marty") Weiss, a Times Mirror Cable network architect, worked from an office on the ASU campus to help bring the EC-Net project together. Part of the project's impetus, he pointed out, was about shifting the perception of what a "cable" company did for a living. Times Mirror was intent on proving that its flagship cable operation in Phoenix—later sold to Cox Communications—could play in the telecommunications big leagues, alongside the regional Bell companies and flashy new fiber builders like the New York–based Teleport Communications Group. "We were looking toward the future," Weiss recalled. "We had no credibility as a carrier at that point. We were still considered 'the cable guys,' so there was a tremendous amount of effort to try to create a new image."

In an early 1994 report, Weiss recounted how the aerospace companies had wandered off on their own to devise clever applications of their new high-speed, long-distance data network. "Our beta test customers are coming up with ways to use the technology which we had not considered in our original plan," Weiss observed.

This was an important point. EC-Net was originally designed by the director of ASU's Computer Information Management Research Center with a narrow mandate: a means to help Arizona tech companies compete for business with Mexico, where low labor costs gave big advantages to companies with facilities there. The center's director theorized that connecting small and medium-sized tech companies with larger peers could supply advantages to Arizona's tech community by speeding up processes around bidding and procuring business. Time, not money, would be the competitive differentiator.

But by March 1994, EC-Net was doing more than speeding up negotiations and payment cycles. Instead, interactive whiteboarding and collaborative design had emerged to provide unexpected benefits by enabling users to send and receive high-resolution electronic designs, warding off ambiguity that sometimes led to manufacturing problems. One user, a vice president from the area manufacturer Tempe Precision, enthused about how quickly computer-aided design processes were happening over EC-Net: Completion cycles for some projects were reduced from weeks to hours. Geography-bending applications like work-from-home telecommuting and live videoconferencing were also popular. Weiss even alluded to a "multimedia warehouse" that enabled users to surface documents, videos, and catalogs. And he wrote about "high-speed access to the Internet for electronic mail." Lynn Jones, a program manager for DEC, shared a memo marveling at how the network was able to scale to accommodate rising usage.

Emboldened by the early findings, Times Mirror extended the EC-Net network to the home front. In nearby Paradise Valley, working on a hunch that there was more to this broadband category than just empowering big companies, Weiss and his team connected around one hundred homes to a two-way, fiber-fed data network, again using LANcity modems. Weiss remembers touring the neighborhood with journalists and talking with early users, including a stay-at-home mom who told him she couldn't believe how fast the Internet had just become. File downloads, in particular, were a focal point for these early adopters. Documents and digital photographs users once labored to download shot across the network in seconds. "File transfer was one of the killer apps," Weiss recalled.

The work caught the attention of Jim Robbins, a highly regarded industry figure who was CEO of Cox Communications, the company that had acquired Times Mirror Cable in June 1994. Robbins maintained his main office at Cox's Atlanta headquarters, but he and a team of colleagues were now showing up regularly in Phoenix, curious about what Weiss and team were brewing up.

If Robbins was interested, it meant the nascent cable-broadband movement was starting to get attention from the CEOs: Individuals like Robert Miron, the statesmanlike and influential president of Newchannels Corp., and Brian Roberts, the CEO of Comcast who, in the winter of 1996, would wow a crowd at the annual Consumer Electronics Show with a demonstration of a 10-megabits-per-second cable modem. These were individuals who would decide whether to greenlight multibillion-dollar investments in network upgrades. Robbins, especially, would become a LANcity champion and a personal mentor whose pedigree helped our small company build its profile in cable-land. He was an exceptionally perceptive individual. During one meeting with me at Cox's Atlanta headquarters, he pointed to a Persian rug decorating his office floor. The rug's intricate design was created from centuries-old patterns prescribed by exacting numbers of knots per square inch. It was a reminder, Robbins counseled me, to be patient. The work we were undertaking was going to take time.

Robbins and his management team weren't the only high-powered people starting to take notice. The work Times Mirror and Cox were doing in Phoenix also caught the attention of the Clinton administration and, in particular, the interest of Albert Gore, the nation's vice president. I was among a few dozen people who had been invited to the White House in the summer of 1993 to demonstrate the possibilities of cable modem technology. Times Mirror Cable Television's president and CEO Larry Wangberg, by now a fast friend thanks to our collaboration in Phoenix, wanted me there in the event that a technical question might arise. Sure enough, it did. After handing over our IDs, making our way through a gauntlet of metal detectors, surviving two head-to-toe body searches, and being admonished not to pet a serious-looking German Shepherd, I walked with Wangberg and his entourage to the Rose Garden bordering the Oval Office and the White House's West Wing. Standing before our group was Gore, who seemed captivated by what we were displaying: a telecommunications network that could make a tedious Internet fast—and bring the superhighway vision that much closer to reality.

Gore was a tech enthusiast—one of the so-called Atari Democrats. As a senator from Tennessee, he had championed the High Performance Computing Act (in December 1991), which sought to compel improved standards for connected computing while retaining the decentralized quality of the Internet. Gore's bill plowed $600 million into initiatives like the National Information Infrastructure and funded the National Research and Education Network, both of which furthered efforts to inaugurate a digitally connected version of America. Now, he wanted to know: Just how fast was our cable modem?

A good question. Wangberg and his team had been talking up the wonders of the EC-Net project, but the demonstration being displayed in the Rose Garden wasn't actually live; the District of Columbia cable system that pumped signals to the White House hadn't yet been upgraded to accommodate high-speed data delivery. Instead, Gore was shown a videotaped simulation of what had been going on in Phoenix: Engineers from multiple locations collaborating on three-dimensional designs and communicating via videoconference.

I waited silently for somebody to respond to Gore's question. But none of the VIPs gathered near Gore seemed to have a ready answer. Wangberg deferred to a colleague standing behind him, who turned and deflected to a second invitee. By the time the game of pass-the-question extended to a third, I decided to resolve the suspense.

I raised my hand. "Mr. Vice President. Thank you for asking," I said to the nation's second-in-command. "The answer is ten megabits per second, running over a 6 megahertz cable TV channel." Gore nodded in acknowledgment, seeming to understand that 10 megabits was an impressive feat. Mission accomplished.

* * *

This scenario was a kicking-off point for a category that would rise to become the no. 1 revenue contributor to the cable industry. Cable television, a business that had revolved around the "television" part of the equation since its late-1940s inception, was about to begin a long march toward a new model, where more money would come from high-speed Internet connectivity than from video subscriptions.

But nobody knew that yet. Instead, another promising idea was gaining steam: delivering telephone calls over cable lines. Some industry executives were convinced that telephone service, not the newborn category of data connectivity, would be the ticket to revival. When CableLabs issued a request for proposals tied to cable-phone technologies in the summer of 1994, the organization's president and CEO, Richard Green, called it "the best thing that's happened to cable since digital video compression." With residential phone bills averaging $40 or $50 per month, it wasn't hard to calculate the opportunity to seize a slice of the residential voice marketplace. Broadband, meantime, was still mostly unknown.

Because of that, cable modems remained a few rungs down on the business priority list. But at least we were getting noticed. By putting LANcity's cable modems to the early test in places like Castro Valley, Phoenix, and Cambridge, the nation's cable companies were beginning to get a sense of what their future could look like. David Fellows had become a big LANcity supporter, excited by what he saw as a budding opportunity to transform Continental's place in the telecommunications universe. He was also a clever businessperson: Fellows was one of the first people to suggest that we make provisions in our technology stack to allow for varying data rates. That way, rather than offer a one-size-fits-all service at 10 megabits per second, cable companies would have latitude to "tier" their data offerings, applying different price points to different speeds—a smart, ahead-of-its-time idea.

Our development philosophy helped make this possible. We had decided early on that it would be a good idea to build programmable software modules that would enable exactly the sort of data-rate manipulation Fellows was suggesting. Call it intuition, or a lucky hunch. But their forward-thinking architecture work, along with similar future-proofing determination of our software team, enabled latitude in the way services could be offered.

At the same time, the cable industry was starting to carve out notoriety in the broader high-tech world. During the 1994 Western Show, convention organizers and a team from CableLabs had conceived of a dedicated pavilion that would demonstrate the new prowess of the modern cable television system. Big-name companies like Eastman Kodak, the fiber-optics manufacturer Corning, and others were invited to band together to show off capabilities like video conferencing, high-speed data transmission, and online access.

The pavilion, dubbed CableNET by its organizers, CableLabs and the California Cable & Telecommunications Association, was a sort of technology smorgasbord, a coming-out party for the new possibilities of revved-up, highly advanced cable systems. The America Online people were there. So were folks from Prodigy and CompuServe, along with pretty much anybody that developed or sold technology requiring a networked data environment, including LANcity. The message we all wanted to convey was that cable was shedding its TV-centric roots and that big tech should take notice. "We were trying to show big corporations that the cable industry wasn't just this climb-a-ladder, kick the tire kind of outfit," recalled Michael Schwartz, a CableLabs vice president who helped to organize the demo. "We wanted these companies to know our networks had a lot of capability that could support their technologies and their applications."

As momentum rose, I took stock. With more than six years of development work behind us, I was convinced we were ahead of the other guys—big tech companies like General Instrument, Hewlett-Packard, Scientific-Atlanta, and Zenith Electronics, along with newcomers like Com21 and Terayon Systems. For years now I'd been touting the merits of high-speed data over cable lines to almost anyone who would listen: journalists, industry engineers, SCTE meeting attendees, international customers. My sermonizing captured the interest of a well-known industry technology analyst, Leslie Ellis. "From the earliest days of high-speed," she wrote, "Yassini would arrive early and stay late in trade-show press rooms, snaring reporters and analysts to brief them not only on LANcity's progress but the industry's. His frequent mention of the 'million-node network' went far to educate the observing communities about the potential, and achievable, scale of high-speed data."

True enough. The "million-node network" was a major LANcity theme. More than selling modems, my goal was to wake up an entire industry to a business proposition that I had been certain, since my early days at GE and Proteon, would change the world for the better. I'd become entranced by the idea of broadband as a sort of alphabet—a common language everyday people could use to engage with people anywhere, to make their mark. Broadband could be the trigger technology for a new economic construct, where consumers become producers on a global scale, where intermediaries no longer make the rules, and where good ideas can come to life for ordinary, everyday citizens. A one-million-user, fast data network running over distances of up to 200 miles had never been accomplished. Now it was within reach.

CHAPTER 11

WICKED FAST

The vaunted "convergence" of communications sectors, long dreamed of and much delayed, was beginning to take shape.

The dove had flown. The tilt scare had been subdued. Our new chips were working. Thanks to the miracle of silicon, we'd reduced both the form factor and the cost of our next-generation LCB modems dramatically. Support from DEC allowed us to design and manufacture a powerful device being used by healthcare providers, city governments, universities, and more. Our list of paying customers in the cable industry was steadily growing. Our SCTE friends loved our puppies and wanted more of them.

Now, our new focus was leapfrogging to a new personal cable modem—smaller and more agile and affordable for everyday consumers. We would call it our LCP—with the "P" standing for "personal." We were certain it would ignite the residential broadband-over-cable category.

Again, hurdles rose. Our cable industry friends had made it clear that to even sniff the idea of a mass market, we'd have to slash the price of our residential modem dramatically. Our list price for our third-generation LCP cable modem was $499—one-tenth of the LCB modem's $5,000 price tag. We had to remove a space-consuming fan but still find a way to keep the box cool. We had to shrink the form factor for mass-market adoption. Volume manufacturing was another big focus. We had to scale up an automated manufacturing process to be able to produce 1,000 LCP modems in a day. I recruited a former Proteon colleague, Bob Riano, to help drive this much larger manufacturing operation.

We got to work. Even before our second-generation modem was in our hands, we'd started design work on the LCP. I called on our dynamic trio of John Ulm, Kurt Baty, and Bill Corley to engineer our silicon architecture, aiming to complete our development work in January 1995. Our new chip was a 155,000-gate integrated circuit that encapsulated the key contributions of our QPSK modem: a mélange of customized digital signal processing functions and a revamped protocol, including remedies for frequency, amplitude, and delay distortions—an outcome from our adventures in Stow. Remarkably, the powerhouse chip and surrounding components could fit neatly atop a single six-inch-by-nine-inch circuit board.

While the silicon work was still underway, we pored over physical design possibilities for the new slimmer-profile LCP modem. Mike Sperry had come up with an inventive succession of protruding metal flares that would help to dissipate heat generated by the fierce interplay of circuitry underneath. No more noisy, space-hogging fans.

I'd taped pictures of Mike's conceptual LCP designs to a wall at our office, inviting team members to chime in. My goal wasn't to take a vote so much as to gauge reaction. Mostly, I was interested in knowing which design would compel the greatest amount of interest—positive or negative. Our physical design, with its protruding grid resembling the radiator of an automobile, would become the signature of LANcity. We managed to stuff everything into a metal chassis measuring 2.6 inches by 6.6 inches by 10 inches. The new LCP was designed to either stand on one end or to lie flat on a surface; cable company partners had told us they wanted to be able to use either orientation.

We made huge leaps in terms of capabilities. The LCB was a hybrid of analog components and digital circuitry. With the LCP, we went all-in on digital, swapping out analog signal processing for digital signal processing and, in doing so, dramatically compressing the form factor. We'd also widened the field of play for our cable industry partners. We were still aligned with the 6 MHz channel scheme, but now our modem could operate anywhere up to 750 MHz, giving cable companies more leeway to find available spectrum. The LCP also had a wider range, capable of supporting round-trip distances of up to 200 miles.

Excitement was building as I hired more employees and peered around the corner to a promising future. We'd gone from shipping one 80-pound dinosaur per day, with my colleague Gene O'Neill dutifully aligning components, to prepping and manufacturing 1,000 personal cable modems in the same time frame. We were on the cusp of even greater scale economies, within striking distance of a once-unimaginable list price of $299.

Enthusiasm was building in the outside world, too. You could feel it. Almost every week, one of the mainstay trade publications for the cable industry featured an article about some cable company, somewhere, extolling the promise of high-speed data delivery. We'd come a long way from the vibe a few years before when a senior TCI executive had politely suggested Les Borden and I were at the wrong industry convention. A combination of competitive pressure and economic ambition was redefining the cable industry before our eyes.

High-speed Internet service was on the mind of the telcos, too. A transmission technology that went by the acronym DSL, for "digital subscriber line," was close to being commercially introduced by the Bell companies as a way to try to forestall what they feared would be a serious incursion by cable television into the US Internet delivery marketplace. DSL faced constraints on distance and speed, but it still loomed as a serious competitive force.

Essentially, this meant that the vaunted "convergence" of communications sectors, long dreamed of and much delayed, was beginning to take shape. Forget the failed videotex ambitions of the past or the halts and starts around interactive television. It was becoming clear that a confluence of powerful PCs, enticing web browsers, and fortified high-speed data lines was conspiring to make the world a very different place. In scattered deployments around the country, cable modems were starting to make their presence felt. Prominent cable providers, including Viacom, TCI, Continental Cablevision, Media General, Comcast, Jones Intercable, and Cox Communications, were installing modems in selected "friendly" houses, usually those of employees and business partners, across multiple markets. Our modems powered most of these implementations. In at least one case, we'd displaced another competitor whose technology, I was told in confidence, had failed.

Here again, we enjoyed a head start. For some of our competitors, these were the first real-world deployments. Not for LANcity: We'd already been making inroads from the outside-in by doing business with international cable customers and smaller US operators. England was an especially important proving ground. In London, the Telewest partnership of TCI and US West was sinking money into an ambitious new cable system conceived to take on the giants of UK telecommunications. One was Britain's dominant phone service provider, British Telecom. The other competitor was the media mogul Rupert Murdoch's BSkyB, a satellite television company that dominated the pay television market.

It was the first ambition—giving British Telecom a run for its money in the telephone business—that inspired Telewest to go big from day one with a modernized, two-way cable architecture. The Telewest network featured deep fiber runs, fine-grained neighborhood serving nodes, and a pristine reverse path. Translation: "It was just wide open for cable modems and the Internet," recalled Chuck Carroll, US West engineering specialist who traveled to London in 1989 to take on the Telewest buildout program.

Carroll was there to oversee the laborious effort to build out a pristine "newbuild" cable system in a dense, urban environment. On the ground, work crews jackhammered their way under London streets to lay fiber conduits, an expensive proposition but one that thrilled Carroll. "We were creating the kind of ecosystem you could start playing around with," Carroll recalled.

Carroll was watching the opportunity take shape from two vantage points: as a telecommunications engineer and a Seattle Mariners fan. From his London flat, the US expat had been going online via a phone-line service from a local provider, Videotron. A main allure was keeping up with the daily exploits of Mariners stars like Ken Griffey Jr. and the southpaw pitcher Randy Johnson. The experience of "going online" made Carroll an early believer. "You just knew there was something there," he said. "I remember looking at it and thinking: This is the future."

I loved working with Telewest and its overseas cable peers because of their appealing management approach. The layers of management and the bureaucratic impulses of large US cable companies were nowhere to be found. Completing purchase orders with companies like Amsterdam 2000, a Netherlands cable company that had embraced LANcity technology, took days, not months.

Life was different domestically. For the most part, we were still in the very early stages of the cable modem growth curve, delivering a few dozen modems here and a few dozen there. At Viacom Cable's advanced-services test bed in Castro Valley, a few dozen homes had been outfitted with LANcity modems by early 1995. Top cable industry executives still wanted to proceed somewhat cautiously. As a result, much of the world had little clue what was happening. *PC Magazine*, a popular hobbyist journal, was still running reviews of dial-up modems.

But that was about to change.

* * *

In the spring of 1995, I was managing a juggling act familiar to any aspiring entrepreneur: how to prepare for growth without disrupting the positive chemistry and culture we had baked into our company. Being able to fulfill orders meant we had to have inventory on the shelf, ready to ship. No cable company was going to wait six or nine months for their orders to materialize. Because of this, critical components like flash memory chips and surface acoustic wave filters had to be preordered—and paid for—nine months in advance. Earlier in the year, on the cusp of a critical deadline, we'd survived a major scare when the chipmaker Intel, with no advance notice, canceled work on a promised flash-memory plug-in. I panicked, knowing the delay could freeze our manufacturing efforts in midstep. Salvation arrived when familiar colleagues from Intel's rival Advanced Micro Devices (AMD) told me they could supply us with an alternative flash memory chip. One more huge sigh of relief, and one more indication of why personal relationships mattered.

Our circuit boards were assembled by a Westborough, Massachusetts, company, Design Circuits Inc. Boxes of finished boards would arrive at our shipping dock, where we'd affix them to our metal chassis, put them through a rigorous testing regimen, and ship them in sturdy cartons to customers, such as Australian companies Optusvision and TeleAustralia and UK's Telewest. We were also starting to see interest from the bigger North American cable companies. By September, we were doing business with nearly a dozen prominent cable providers. Our growing company was now doing business with nine of the companies on the US cable "top 10" list.

In Canada, we were building momentum, too. Shaw Communications, Cogeco, and Videotron, all considered industry bellwethers with strong technology credentials, were placing orders. So was Rogers Cable, whose principal architect, George Hart, was among our most demanding partners, insisting on rigorous technical analytics before putting our modems to use.

Rising modem purchases translated to rising revenue for LANcity. Back in 1992, our revenue hovered around $2 million annually, nearly all of it flowing from service contracts and almost none of it coming from product sales. By 1995, we were pacing around $6 million, with our second-generation LCB modem accounting for most of the revenue and our third-generation residential modem starting to move off the shelf. Our net earnings were still nothing special, but bottom-line profit wasn't the point. We wanted to establish a beachhead as the market share leader, with top-line revenue momentum as our primary financial goal. I was plowing almost all the revenue back into the company by hiring more operations personnel, developers, and engineers and by automating our manufacturing. LANcity was in business.

By now I could feel it: The cable-broadband vision we had set into motion in 1987 with a sheet of paper and a list of key features was going to explode. Researchers from the Kagan organization had run the numbers, estimating that in 1994 there were nearly 60 million US homes connected to cable television systems. In 24.7 million of these homes—slightly more than 40 percent—there was at least one computer whirring away. That number was expected to swell to 41 million by the end of 1998. The vaunted "convergence" model was taking shape right in front of our eyes: souped-up cable television systems, upgraded to accommodate two-way data delivery; an impressive rise of the home computer; and LANcity and a handful of cable modem makers figuring out how to transform cable's fat pipes into vessels for high-speed Internet traffic—at a sweeping scale both in terms of users and of distances covered.

Even so, the big prize—a mass-market, thriving consumer broadband sector—was barely born. We were the runaway leader in the cable modem category, but so far, we had shipped only about 20,000 modems—more than any other manufacturer but not enough to signal a revolution. Instead, a newcomer in the cable-broadband category would propel the market forward, a former Intel sales executive turned investment banker named John Doerr. During the cable industry's December 1994 Western Show, the Silicon Valley investment advisor had happened upon a demonstration of a cable modem. Doerr, already a tech industry legend, had led his firm, Kleiner, Perkins, Caulfield, and Byers, in early-stage deals involving home-run companies like Sun Microsystems.

Doerr was strolling around the CableNET pavilion put together by CableLabs and the California Cable & Telecommunications Association. Eyeing a cable modem connected to a live cable network, he was seeing for the first time the missing link in the whole information-highway equation. Impressed by what he had seen, the Silicon Valley venture capitalist worked feverishly in the first few months of 1995 to assemble a business venture led by TCI and funded by Doerr's firm. The partners adopted a hip-sounding name for their venture: @Home Networks.

The @Home venture started out humbly. When I first visited the original office near San Jose, California, there was only the barest evidence of an operating company: a few rented desks and a rack of blinking equipment. But the action heated up quickly as Doerr recruited William Randolph Hearst III—a longtime newspaper editor and an executive with Hearst Corp. (and the grandson of publishing mogul William Randolph Hearst)—as CEO. Besides his newspaper heritage, the younger Hearst brought a cable industry pedigree to the role, having previously been a vice president for Hearst's cable television division.

@Home was not exactly the cable hive-mind personified, but it came close. After attracting Comcast and Cox Communications as equity partners, @Home was owned by partners that collectively represented close to half of the cable subscribers in the United States. The main holdouts were Continental Cablevision, which had separate linkages with a large Internet service provider, Performance Systems International, and Time Warner Cable, which launched a like-minded entity called Road Runner. Either way—@Home or Road Runner—the ambition was identical: bring scale, purpose, technical know-how, and energy to the cable-online opportunity.

Handing these duties off to a newly formed company was a telling decision. Behind @Home's creation was the belief that rank-and-file cable companies, historically providers of television channels, were ill-prepared to take on the task of introducing and provisioning a brand-new, very different sort of data communications service all by themselves. Instead, @Home would do the heavy lifting: constructing and operating a high-speed backbone network and assisting cable companies with the arduous job of "provisioning," which involved integrating the newly arrived high-speed Internet business with existing back-office billing systems that were meant to accommodate simpler pay-TV services. It was a vexing task, requiring that a cable company be able to sniff out the unique address of a connected cable modem in order to assign a monthly invoice to a particular household. To help guide the complexities on the technology front, @Home hired former NASA scientist Milo Medina, a highly regarded engineer, as chief technology officer.

Beyond technology, @Home was also concerned with what appeared on the screen, creating a business arm devoted to web pages, applications, games, newsfeeds, chat rooms, and the like. @Home went so far as to enlist the services of a well-known magazine designer, Roger Black, to devise a home page that would carry a polished, professional look—more akin to a modern magazine layout than the stodgy feel of early-era web design.

@Home seemed to be following the AOL model: providing not just a means of connecting over wires but a repository of content that would be available exclusively to @Home subscribers. The cable company itself, whether it was Comcast or Cox or TCI, would yield the consumer-facing stage to @Home—a sexier, modernized offshoot. The reasoning was that nobody in the cable business quite yet understood the role the industry was supposed to play in delivering a service that looked to be very different than the familiar cable television offering.

Although cable modems were a vital enabler for @Home, for our part, we weren't so sure about the @Home business model. Doerr and his team believed that what consumers wanted was a do-it-all aggregator: a "fat pipe" coupled with an exclusive pool of web content and plenty of interesting things to do once the connection was made. This pairing of distribution and content was torn straight from the cable industry's decades-old television model, where customers paid a single fee both for connectivity and for television content. AOL had the same ambition, having stitched together content with phone-line connections underneath one service invoice from the very start.

Regardless of the business construct, @Home's contribution was significant. The involvement of heavyweight cable companies, the participation of a well-regarded Silicon Valley investment firm, and the energy brought to bear by @Home combined to energize the emerging cable broadband category. It also delivered big worries to AOL: As @Home rose up as a potential competitor, AOL tried hard to get into the high-speed, cable broadband game, angling for alliances with major cable companies that could blunt @Home's inroads, to little lasting effect.

@Home also provided something the cable industry did not possess until now—cachet with Silicon Valley. @Home's new headquarters wasn't tucked away in cable industry strongholds like Denver or Philadelphia but instead had been moved to 450 Broadway St. in Redwood City, in the heart of the valley. Once there, @Home added street credibility in the form of a well-regarded successor CEO—a tallish, tousle-haired surfing enthusiast named Tom Jermoluk, who previously held a senior position at computer manufacturer Silicon Graphics.

The Valley ties were important: They signaled the stodgy, pole-climber heritage of the cable industry was morphing into something new. It was an important change of attitude that would usher in a series of major alliances uniting the two camps. This included Microsoft's $1 billion investment in Comcast in June 1997, a deal that almost single-handedly restored positive perceptions around what had been an ailing cable industry.

The Microsoft vote of confidence told the world what we already knew: that the key to the information highway was the cable industry and the high-capacity networks it had built across more than 90 percent of the United States. It was a precursor to tech maven Paul Allen's acquisition of a controlling interest in the cable companies Marcus Communications and Charter Communications and, later, the behemoth of them all, AOL's January 2000 acquisition of Time Warner. @Home itself would get into the deals game, too. Its 1999 acquisition of Excite, then a prominent web portal competing with the likes of Yahoo!, deepened the cable-meets-tech ecosystem. Here was a company owned by old-school cable guys, buying its way deep into the emerging digital media world. It was the start of a generational migration for the cable industry, from a hardware-centric model to a software and services orientation.

@Home would grow up fast. The company's residential subscriber total nearly doubled during the first quarter of 1998, reaching 90,000 and representing the large majority of the nation's total cable modem users. Subscribers bought high-speed Internet service under the @Home brand, with the local cable company providing the underlying pipe, connecting the cable modems inside homes, sending out the monthly bills, and splitting the revenue.

Time Warner's Road Runner Internet service also picked up steam, racking up 40,000 subscribers by March 1998. Another allied offering, MediaOne's Express service, had an additional 30,000.

In the end, @Home would unravel under the weight of conflicting partner agendas, the late-1990s dot.com bust, and boardroom jousting involving the TCI successor AT&T Broadband and its cable partners. By 2001, after several changes of direction, a decaying balance sheet, the pricey acquisition of the online portal Excite, and more corporate drama than anyone had ever imagined, @Home filed for bankruptcy. Its intra-industry cousin, Time Warner Cable's Road Runner, also would retreat from the scene, partly because of the theatrics associated with another megamedia merger, the star-crossed combination of AOL and Time Warner.

But the broadband-over-cable category would shrug off these external machinations. Cable companies, including those that had invested in @Home, began to realize they could get by just fine without the brotherly support of any external enabling agent, so long as they adhered to the familiar duo of standards LANcity had long embraced: Ethernet for data transmission and 6 MHz TV channels for the pathway to the home. Early @Home backers like TCI (and subsequently, AT&T Broadband), Comcast, and Cox Communications peeled away from @Home, shifting responsibility for the entire high-speed Internet business to internal teams. A "war room" AT&T Broadband established to oversee the transfer of cable modem service from @Home in the course of just a few days would become the stuff of industry legend as a resolute TCI technology executive, Susan Marshall, commanded an around-the-clock effort to help the company's systems get provisioned at breakneck speed, mostly over the course of one very long, strenuous weekend. I marveled at how Marshall and her team seemed to have accomplished a miracle, the equivalent of tuning up a car engine while screaming down the highway at 100 miles per hour. When it was completed, the project underscored the new reality: The demand for high-speed Internet service over cable would be enough, in its own right, to catapult the business forward. That much was underscored by a customer quote shared by Doug Semon, the Viacom Cable engineer who was among the first, if not the first, to install residential cable modems for paying customers. "You can have my cable modem," went the oft-cited quote, "when you pry it from my cold, dead fingers."

* * *

With DEC at our side, we were going places. Under the banner of "ChannelWorks," DEC's team joined us to storm the cable television industry, bringing impressive data networking credibility plus marketing know-how and prowess to our work. We went from LANcity's humble tabletop displays at industry conferences to neon-lit booths where DEC and LANcity together paraded our modem technology. At these cable industry conventions, we made for an unfamiliar sight: Amid brightly lit booths populated by the likes of Home Box Office and ESPN, here was a top technology company touting the marvels of a cable modem that could help the industry usher in an entirely new business.

Looking over the crowds at the cable industry's two big conferences—one held annually in Anaheim, California, the other moving from city to city—my DEC colleague Jerry Amante could sense something big was about to happen. Amante was certain our work was going to be groundbreaking, for two reasons. First was the understanding that our advanced modem platform did what we designed it to do: transfer data over residential cable systems at multi-megabit-per-second speeds. Second was the enormous presence of that same cable television infrastructure, with lines lit up across more than 90 percent of the nation's homes.

Jerry Amante would end up devoting nearly four years of his career to the data-over-cable pursuit, carefully evaluating our business progress, meeting with our customers and our prospects, and caucusing with DEC colleagues at regular intervals for strategy and design reviews. Amante's abiding question was whether LANcity was holding up its end of the bargain in order to continue receiving injections of development capital. "It was all about: does LANcity deserve its money?" Amante would recall three decades later. At every major milestone, the answer was "yes." And thankfully so. Without DEC's funding, our efforts to develop a wholesale revamp of our modem technology would have been dead in the techno-water.

But other agendas would short-circuit the dream for DEC. After allying with us since late 1992 as an OEM partner and offering informal counsel to me since 1989, a revered leader in the Ethernet space had a decision to make. Per our agreements tied to the development of the second-generation LCB modem, DEC also had first-refusal rights on the latest iteration of our technology, our third-gen, mass-market, and residential cable modem. Beyond that, DEC could take our relationship to the next level by investing directly in LANcity or, potentially, even buying our small company outright. I had presumed option no. 2—making an equity investment in LANcity—would win out. Why not? A tech giant that was refashioning itself as the king of networked computing seemed poised to emerge as a powerhouse player within the category of metro-wide, high-speed networking using residential cable infrastructure.

That wasn't all. Thanks to the partnership with LANcity, DEC already had a clear first-mover advantage. At the time our second-generation LCB modem was on the market, nobody had achieved as much as we had with a data-over-cable technology suite that was suitable for deployment—capable of correcting on the fly for various network conditions and enabling self-installations and intelligent self-management. We had spent years developing a digital implementation of the QPSK modulation scheme that obviated the need for manual tuning and was now entwined in a graceful dance with our brand new MAC protocol. More importantly, we were adapting our knowledge and development chops to produce a third-generation modem that was poised to dominate the biggest market of all: the residential high-speed Internet sector.

But it was not to be. At DEC, the early involvement with cable modem technology had dovetailed with internal strife. The company's inability to reckon with the personal computer revolution had sent DEC into a tailspin. A loss of $2.8 billion for the fiscal year 1992 was one of the largest financial tumbles yet in the computer industry's history. In July 1992, DEC's founder Ken Olsen, the man who had started the company in 1957 with a $70,000 loan, surprised company watchers by announcing his retirement. His successor, Robert Palmer, shared little of his predecessor's zeal for being first in line with breakthrough technology. The focus instead was on the profitability of existing business lines.

Even though our technology showed promise, the one hundred or so first-generation LANcity modems DEC had ordered weren't moneymakers, at least not yet. The idea was to use these products as loaners and starter kits for DEC's internal sales offices, to establish a beachhead of knowledge and understanding. DEC recognized there would be an elongated ramp-up period for the high-speed cable modem business. To that point, some DEC insiders had been referring to our modem as the "camel's nose"—shorthand for a product that served as an entry point more so than an immediate profit center. But that approach ran counter to the norm for the giant tech company. There was also contention for resources. The business unit DEC established to market our cable modem technology, headed by a rising executive, Bill Styslinger, was charged with developing a broad portfolio of networking products, including a digital video advertising system with strong appeal to a cable industry hungry for immediate revenue streams. Our modem was just one in a long line of products that needed care and tending.

When Amante called to schedule a face-to-face meeting in December 1995, I had a feeling something was wrong. When he arrived, he seemed unusually somber. He'd driven over to tell me in person that the final decision was a hard and nonnegotiable "no." No investment in LANcity, no continuation of our OEM agreement, no exercising of the first right-of-refusal for our residential modem. "It's nothing you did wrong," he assured me. "It's just internal stuff happening."

I had been working side by side with Jim Albrycht and John Kaufmann since 1989 to further our shared ambitions in what Albrycht liked to call the "community multimedia networking" space. Now, our DEC partnership was over.

But its impact would linger. Our alliance with the larger-than-life tech company had provided us with $2.5 million in funding to pursue the vision we'd discussed in Mary Lautman's living room years before. Without DEC's endorsement and partnership, we never would have gotten this far, this fast.

I was certain DEC could have dominated the broadband-over-cable category on a huge scale by furthering the alliance with LANcity. But the company that had been determined to play a lead role in metro-area networking—the "Ethernet over Television" champion—had backed away.

For some people at DEC, the withdrawal was disappointing. Amante had advocated for further involvement with our company, having seen the possibilities up close after hitting the road to visit with a handful of our customers: the wastewater department of the city of Beaverton, Oregon; the Milpitas, California, school district; and others. He'd been impressed after seeing one of the first digital "white board" applications, with team members from various Arizona State University buildings and corporate partners able to swap ideas and schematics in real-time across a novel screensharing environment. In the background and with little recognition, DEC played an oversized and under-appreciated role in developing the modern information superhighway.

Amante would remain a close friend. He was disappointed about DEC's withdrawal, but not regretful. To this day, he considers his work with LANcity the high point of a career that spanned decades in advanced technology fields. "It didn't help DEC in the end," he said about the DEC-LANcity alliance. "But it helped humanity."

* * *

With modem fever building, cable industry leaders were determined not to invite a repeat of a technology split that was the enemy of unity and scale. For decades, cable companies had grappled with a bifurcated model, where a pair of dueling technology platforms from two different manufacturers, General Instrument Corp. and Scientific-Atlanta Inc., were incompatible with one another. The industry was effectively cut in half from a technical standpoint, making it difficult to merge operations or achieve important scale economies or keep up with the rapid pace of innovation in the computing industry.

A turning point had begun to take shape late in 1995, just before DEC exited its partnership with LANcity, as the cable industry technology consortium CableLabs began to train its focus on cable modem interoperability. The idea here was to avoid a second technology schism and instead ensure that different cable modems from different vendors would adhere to a common technical foundation. It was the start of a lengthy wave of independent tests that would usher in a global standard for cable modems.

The initiative would go by the unwieldy acronym of DOCSIS—for Data over Cable Services Interface Specification. It was designed to follow a path familiar to the broader technology and computing industries, where standardization of components and systems had led to important economic and consumer benefits. One example among many is a standardized receptacle for electrical appliances and cords. Imagine, otherwise, rooting through myriad combinations to find a table lamp that would align with one of many incompatible wall outlets.

Part of the interoperability quest was rooted in the dream of developing a retail market for cable modems, where consumers could select a device off the shelf at their local Best Buy store with the assurance it would work with their cable company's broadband service. There was also growing confidence that cultivating a broad pool of modem makers would help to bring costs down as volume manufacturing swelled.

CableLabs prepared a request-for-information document involving a lengthy checklist of attributes, with modem vendors, including LANcity, asked to answer questions about technical specifications, development history, future plans, and more. Tucked into the CableLabs request-for-information document was one of the key queries that would end up unleashing the DOCSIS objective. It dealt with the respondent's approach around media access control—a foundational protocol governing how a modem handles data transmitted over a long-distance cable network. CableLabs was asking a series of questions about how modem makers would approach this critical piece of the puzzle.

It was terrain we knew well, having worked since 1988 to devise our company's Unilink-II protocol. A US patent submitted in June 1993 by five LANcity developers captured the breakthroughs behind our work, describing the system as "high-performance, high-volume communications highway network system" and referencing the myriad of workarounds and solutions our team had come up with to resolve the intricacies of metro-wide data networking.

I thought about the patent filing and the amount of determination behind it as I looked over the CableLabs document. I was struck by what looked to be one of the biggest questions of all: Was the respondent willing to make its technology available to the DOCSIS initiative, free of any royalties or licensing fees? It was a consequential ask. Much of the software business derives its economic lifeblood from licenses—often expressed as per-user fees that attach to the use of intellectual property. Our MAC protocol was a signature element of our cable modem, effectively its nerve center. Even a modest royalty had the potential to deliver big revenue. It wasn't a stretch to conceive that within just a few years, millions of homes would be outfitted with cable modems. Over time, as the technology's footprint grew, it was conceivable that hundreds of millions of people globally would be connected to these new devices. Maybe one day, billions.

In the LANcity MAC protocol, we knew we possessed something not only special but also future-proof. We had devised the protocol to be malleable, baking in structural logic that would enable broader functionality to be added over time via software upgrade. I thought about the concept behind a popular toy: interlocking Lego bricks that shared a modular architecture inviting endless customization. Lego sets allowed users to create fantastic buildings and shapes from pieces of formed plastic. Our protocol would allow future developers to rev up data rates, add security features, or enable new diagnostic feats.

I thought about the incalculable effort and brainpower our LANcity crew had poured into Unilink-II. I thought about my colleague Cindy Mazza yowling with delight when that first packet passed through our next-generation modem. I flashed on the moment at Proteon when my idea of combining LAN technology with cable television infrastructure was dismissed by an angry founder. That seemed like a long time ago. But my vision had not wavered. I was here to make a contribution. To put my own ding in the universe. More than 30 years later, in an induction address for my entry into the Syndeo Institute's Cable Hall of Fame, the veteran telecom industry technologist Balan Nair would point to this same moment. "That MAC protocol?" said Nair, the president and CEO of Liberty Latin America, "He contributed it to the cable industry at no charge, with no royalties, for perpetuity."

That I did, and in the end, it wasn't a hard decision. When it came time to codify the pledge, I flashed onto a term I'd learned while admiring the play of Robert Parish, the Boston Celtics center whose tough-minded work ethic had inspired the team for 14 years, through the 1994 season. My answer was, in basketball parlance, a layup. An easy call. I checked the box that indicated "yes." Right there on the spot, I gave it away.

* * *

On April 18, 1996, the journalist Walter Mossberg published his latest column in the *Wall Street Journal*. His Personal Technology articles, appearing on the first page of the newspaper's second section, were widely read by technology and consumer electronics enthusiasts who closely followed his to-the-point, often caustic critiques of everything from personal digital assistant devices to new laptop computers.

The object of Mossberg's scrutiny this time was a newfangled entrant on the burgeoning online scene: the cable modem. A few months before, Mossberg had visited the offices of the cable company Jones Intercable in Alexandria, Virginia, where managers had set up a side-by-side display of two computers. One was connected to a phone line via a 28.8 kilobits-per-second phone modem. The other was connected to our new, third-generation residential cable modem, which had recently been introduced commercially at the Jones flagship system. The verdict was encapsulated in a line that would become immortalized in cable industry lore: "I recently tried out the service, called the Jones Internet Channel," wrote Mossberg, "and it is wicked fast."

Wicked fast. It was exactly what the cable industry, weary of pressure on its core video business and anxious to resume a growth posture, wanted to hear. The review was especially convincing coming from the *Journal*, a newspaper religiously read by Wall Street investment strategists who aspired to be part of the Next Big Thing.

Behind the scenes, the makings of Mossberg's prophetic praise were set into motion by a cable industry public relations operative, Jim Carlson. Several months before, the Jones Intercable vice president had invited Mossberg to give the Jones cable modem service a try. Carlson figured Mossberg, who lived in nearby Arlington, might be intrigued.

From the start, the day had registered a positive vibe. That morning, Carlson shared a small-world moment with Mossberg as he greeted the *Journal* writer in the parking lot of the Jones building. As the two chatted, Mossberg detected in Carlson's vowel-effusive accent a regional affectation that seemed familiar. Sure enough, the two figured out they'd attended Pilgrim High School in Warwick, Rhode Island. Both were well-steeped in New England vernacular, where the word "wicked" describes the mighty, the hip, and the ultra-impressive. "That was the magic term," Carlson said. "It would have a huge impact for the industry."

Although Mossberg peppered his column with a few caveats—suggesting the $40 per month price tag was a bit spendy and warning readers that not every cable system in the country was yet outfitted to offer fast Internet service—overall the message was glowing. "A cable modem can deliver up to 10 million bits of data per second, nearly 350 times the speed of the fastest standard modem," Mossberg wrote. "What's more, cable modems are always 'on' and waiting for you, like a TV channel, and don't require dialing or placing a call to get going."

Mossberg's column was an inflection point for a category that was about to energize an entire industry, redefining the way "cable television" companies made money and contributing to a stupendous run-up in company and industry valuations. A technology we had been developing at LANcity for nearly eight years, evangelizing at regional SCTE conferences and in trade publication articles, was about to host its coming-out party. "Cable modems take center stage at NCTA convention" was a headline appearing in *Cable World* magazine's May 6, 1996, edition, chronicling big themes in the conversation at the cable industry's annual National Show held in Los Angeles. Decker Anstrom, the NCTA president, pronounced from the stage at the same conference that a new era was near. "We are on the threshold of literally reinventing this industry," he told an approving crowd.

As enthusiasm rose, tech industry competitors went on high alert. Shortly after Mossberg's column hit, I received an unexpected phone call from a senior executive for U.S. Robotics, the big dial-up modem maker based in Chicago. Days later, on a Thursday, a phalanx of U.S. Robotics executives descended on our office in Andover. They'd flown in on a private jet, with determined looks on their faces. One of the leaders of the group, an amiable executive vice president, got right to the point. "We'd like to talk about buying your company," he told me. As in, outright. As in, tomorrow. This was a Thursday afternoon. They wanted to seal the deal on Friday. A printed letter of agreement had already been signed. The offering price was $40 million.

I politely explained that we weren't entirely finished with the work we set out to do, which was true. But privately, I wondered whether a deal with a prominent dial-up modem manufacturer was the right way to further my original vision. I wasn't sure a company whose lifeblood depended on early-generation technology would share our abiding conviction for making something revolutionary happen.

Early the following week, I received a second letter, insisting that we consider the offer. That was enough for me. I turned any further correspondence over to my attorney, David Fine, a partner in the Boston firm of Ropes and

Gray. I asked Fine, who would remain a longtime trusted friend, to graciously but unapologetically convince U.S. Robotics to call off any further conversation.

Even so, it was clear this wouldn't be our last entreaty. Anybody watching could connect the dots: The Internet was exploding. Home computers were becoming everyday appliances. People were increasingly dissatisfied with slow phone-line modems. And at the center of a budding movement, here we were. A smallish Massachusetts company that had figured out a way to harness the power of the residential cable television infrastructure to make the Internet fast. The idea that we might be a desirable prize loomed as obvious.

Not that we were alone, there were competitors stirring. Companies like the newly conceived Terayon Systems of Santa Clara, California, and the consumer electronics giant Zenith Electronics were starting to be mentioned as cable modem contenders. The tech giant General Instrument announced an order for one million modems in the spring of 1996 to three big cable companies: TCI, Comcast, and Time Warner Cable. Even so, we knew we were ahead of the game. I calculated we had at least an 18-month head start, thanks to intensive development work on our media access control protocol, our digital signal processing technology, deep real-world experience, and breakthroughs allowing our modems to provision themselves automatically within a cable distribution network. We had already begun quietly spreading the word about our new cable modem in May 1995, at the cable industry's big National Show in Dallas. In early December, at the Western Show in Anaheim, the boxes had their official coming-out debut for all to see.

With a list price of $499 and volume discounts that could bring the cost to as low as $299, we knew it could meet the rigid economic demands of cable companies that kept close watch over the costs of their gear. Besides the price, our modem delivered important economic benefits to our partner cable companies by virtue of self-installation features—we were determined to make our modem plug-and-play, easier and faster to install into homes than modems made by any other competitor. Similarly, our residential cable modem was small enough—and, without a fan, quiet enough—to nestle comfortably in a home office, kitchen, or family room. It was a far cry from the hulking, noisy beast that, six years earlier, sat on John Kaufmann's living room floor in Stow.

By September 1995, we counted more than 200 cable companies that had purchased at least some number of modems and our companion headend translator. Smaller, independent domestic cable companies were among our early adopters—and our most enthusiastic champions. Some of the names and their locations evoked an earlier era of cable television, which has its roots in rural America. There was Eagle Communications in Kansas. There was Cable

Alabama, an early LANcity believer. In Maine, we shipped modems to Casco Cable, which later would be acquired by the giant Charter Communications. Even the occasional telephone company slipped into our "active client" roster, with BellSouth in Georgia also on the list. The big guys came around, too. In July 1996, we landed our biggest deal yet: Continental Cablevision ordered 50,000 LANcity modems as it planned major service launches in Boston, Chicago, Detroit, and Jacksonville, Florida.

Even the mighty TCI had become smitten. I'd recently entertained a crew of TCI representatives who had descended on our Andover office, dangling a purchase order for an astonishing number: one million modems. One million! For LANcity! It seemed too good to be true. And I quickly deduced that it probably was: Sometimes when you do business with the giants, there are conditions attached. That looked to be the case here. TCI wanted us to sell modems to them on an exclusive basis, a specter that frightened me. I'd heard stories before of gigantic purchase orders that failed to materialize, leaving vendors twisting in the wind. Plus, I knew the reality of the day. I'd been in the field for years pitching cable modems to cable companies of all sizes as I sought to fertilize the market, deployment by deployment. I knew there was no way one million modems could be digested by one company. There just weren't enough finished, two-way, fully provisioned, and fiber-energized systems within the ecosystem—not yet. I was happy to stick to my Johnny Appleseed strategy. Plant now, harvest later.

Besides that, there was an amusing misunderstanding during our meeting. A TCI procurement specialist had looked around our smallish office and asked me, point blank: "Where's the rest of the company?" The answer was that there was no "rest of the company." This was all we had. A quarterback with a reasonably strong arm would take only a couple of throws to project a football from one end of our office to the other. We had gotten to a place of prominence in the cable industry by dint of hard work and perseverance, not by plowing borrowed money into impressive-looking buildings. The crew from TCI left without a deal and got back on their private jet. No regrets on our side.

Still, even without an order from TCI, we were headed toward big things. Based on early purchasing activity and expansion plans my cable CTO friends had shared with me, my calculation was that 1998 would be the breakthrough year, with our company projected to ship more than half-a-million modems.

Not that we didn't have growing pains. I remember a day early in 1996 when our testing revealed that close to 500 modems were failing—or so it appeared. I was badly shaken. Five hundred modems represented a huge amount of inventory and potentially costly shipment delays. To try to understand what had just happened, we turned our cafeteria into our version of an emergency trauma unit, dismantling modems and testing components until we discovered the culprit—a faulty batch of silicon from our chip manufacturer. Happily, the flawed chips had been stamped with a date code, allowing us to identify the bad batch and replace the malfunctioning chips with substitutes I'd purchased and tucked away—just in case of a disaster like this. We were able to correct the error, restore the modems, fulfill the orders, and avoid the irreversible pain of a hardware recall. Enormous credit was due to my sister Pamela, who ran our quality assurance operations with rigorous determination, working to diagnose problems so that in the morning, returning employees would have the data they needed.

The bigger LANcity vibe, though, was all positive. The mood around our office had shifted from one of nervous determination to a candid realization that we were going to make it. Still, the momentum was coupled with reality. By now, it was becoming clear to me that LANcity was going to need, as the actor Roy Scheider famously put it in an iconic movie about an oversized shark, "a bigger boat."

We had developed a technology inspired by a 1987 vision made real by 13 employees working with a team of 6 advisors. We were on the verge of sparking a revolution. But as the TCI crew had discovered, we were small. Too small, perhaps, to mount the sort of all-out, massive deployment that seemed to be needed. If TCI had reservations about our size and staying power in a ramped-up market, so would other cable companies. And the bigger tech companies knew it.

U.S. Robotics hadn't been our first suitor. Earlier, around the time we were readying our LCB modem, I had hosted one of the most revered executives in the cable industry, the gracious cofounder and chairman of the big equipment maker Scientific-Atlanta. Sid Topol, the man who had almost single-handedly sparked the satellite television revolution that gave Ted Turner and Gerry Laybourne the instrument of their success, was captivated by what was happening in the cable-broadband arena. Flanked by a team including the company's CEO, a tall, physically imposing man named Jim McDonald, Topol asked repeatedly what it might take to pry away LANcity.

I liked Topol, who lived nearby in a home alongside the Charles River. But I wasn't sure about the fit. For one thing, we had a serious difference in perspective on technology. We were convinced the TCP/IP protocol would come to rule the way electronic information traveled across the cable pipe, no matter what sort of data was doing the traveling—text, data, phone calls, video streams, whatever. Scientific-Atlanta was wedded to using the cable industry's video transport scheme, based on what we felt were more restrictive MPEG standards. Moreover, McDonald demanded that I personally guarantee a certain minimum number of modem orders. In 1993, we had almost nothing in our bank account, and despite our rising momentum on the development front, I was still worried about making the numbers work. Failing to meet a minimum guarantee would put me on the hook for a commitment I couldn't back. I told McDonald, in all seriousness, that the amount he had been spending to fly a contingent of executives to meet with us several times—on Scientific-Atlanta's private jet—was more than what we had in the bank.

Topol had graciously retreated. But now, the U.S. Robotics offer, and the realization that people like TCI's procurement specialist saw us as a small player, posed a quandary. One route would be to take our company public, preparing an offering of shares that might assign a lofty value to LANcity. An offering document I worked up with two Boston-area venture capital firms, Matrix Partners and Battery Ventures, theorized we could raise money in a latter-stage financing round that would vault us toward a possible public offering. The numbers seemed enticing. Investors love growth stories, and we surely had one. After overcoming hundreds of battles on the way to mass-manufacturing, by the third quarter of 1996, our modems were beginning to fly off shelves and into delivery trucks. Our staff had ballooned, with LANcity now employing 60 full-time people in the manufacturing department alone. We were a rocket ship, forecasting $40 million in revenue per quarter and rising.

The idea of heading down the path toward an initial public stock offering was intriguing. But the whipsaw life of operating as a public company CEO wasn't something I particularly wanted to experience. Plus, the timing was wrong. We needed a longer track record of revenue and cash flow performance to tell a convincing story to investors. I also knew we'd have to continue to nurture the broader ecosystem at least through 1998.

Most of all, I wanted to know that any pathway for LANcity would lead to a point where the world had a genuine global standard for cable modem technology. Only then could we make good on the vision I'd conjured during my Proteon days: Telecommuting. Connected healthcare. Distance learning. Reuniting families separated by geography. Reducing fossil fuel demand.

There was also a human element to consider. We were growing fast, but there were no more hours I could tease out from any given day. I hadn't slept for more than a few hours on any given night in nine years. I'd willingly and intentionally steered away from major life experiences in order to nurture our technology to health as we confronted a long list of barriers. I wasn't married. I didn't have children. I lived and breathed LANcity. Every ounce of effort I possessed was poured into this company to execute my vision. I knew I'd have one chance in my career to make something this big happen.

My core staff, too, was nearing the point of fatigue. I'd promised them, on that evening long ago in Mary Lautman's living room, that together we could achieve something special. And that they'd be rewarded for sticking with me. A complicated mezzanine financing round, followed later by a hoped-for (but not assured) public offering, seemed hazy. My team had financial obligations I'd promised to help them address. I remembered my pledge to John Ulm's wife when he took a flyer on an unknown tech company run by a CEO he'd never heard of.

We huddled around the conference room—me and my closest compatriots: Mike Sperry, Ed O'Connell, Gerry White, Nick Signore, Chris Grobicki, Les Borden, Cindy Mazza, and Mike Durant. "We're exhausted, Rouz," Mike Sperry confided in a meeting I'd arranged to consider our options. "We've done all we can." We agreed that if the right opportunity arose, we'd support an outright sale of our company. We'd leave the IPOs to somebody else.

There was one interesting option that had circulated in the back of my mind. I'd gotten to know Paul Severino, the founder and chairman of a Silicon Valley tech company, Bay Networks, after being invited to speak at one of his company's user conferences in Orlando, Florida. I liked Severino: a straight-talking, no-nonsense New Englander with a conviction for supporting advanced technology. He'd cut his teeth in the tech trade working for a familiar name—Digital Equipment Corp.—in the late 1970s. Bitten by the same networking bug that had transfixed my friends at DEC, he had started a successful data router manufacturer, Wellfleet Communications, in the mid-1980s in Bedford, Massachusetts. Later, Severino combined Wellfleet with Synoptics Communications, a well-regarded computer networking company based in Santa Clara, California, to create Bay Networks. A trade publication, *Network World*, would later name Severino one of the "hot shots" of the 1990s computer networking arena.

Since we first met, I'd kept in touch. Severino watched our company carve out a leading presence in the cable business with a modem that was now infiltrating the fast-changing industry, not to mention getting written up in the *Wall Street Journal*. I trusted Severino. I was willing to share our corporate strategy and the roadmap behind our company's crown jewel technology—along with market trends and sales projections—in confidence. He came away impressed and interested. LANcity presented a straight-line path to prominence in a coveted sector, cable telecommunications. Almost overnight, Bay

Networks could have a seat at the table with the big guys of cable technology—companies like Scientific-Atlanta—along with a brigade of other electronics firms that had made inroads into the category: Zenith, General Instrument, and Hewlett-Packard were among the prominent names. With little hesitation, Severino and his team came up with a number: $59 million, a substantial premium from what U.S. Robotics had dangled not long before.

To oversee a possible deal, Severino dispatched a senior executive, Bruce Sachs, a friend I had gotten to know in my Wakefield days who would later become a partner in the Boston venture firm Charles River Associates. Sachs's own company, Vialogic, had been acquired by Bay Networks several months before, meaning he knew his way around the process. Sachs assigned a team of associates to run due diligence on LANcity, scouring our books, our engineering documentation, and our economic value. James Spindler, the LANcity legal advisor, did nearly the work of an entire law firm at one-third of the cost, and my indefatigable financial ace, Mike Durant, remained with me as an advisor after working in Applitek's accounting group years before. More than once, Mike had credited me with doing the work of three people at once. Now, it was Mike who was working overtime in support of a possible deal.

My core team members were sitting around a conference table as I told them about the Bay Networks offer. The feedback was almost unanimous. They wanted me to take the deal. They had lives to manage, mortgages to pay, college tuition bills looming. For most of them, this was a once-in-a-lifetime opportunity to cash in at a high level. The response articulated by my colleague Ed O'Connell, a LANcity board member, was a variation on the "bird in a hand" philosophy. There was no assurance we'd be courted by another suitor anytime soon with a better offer. Bay Networks was a respected, innovative tech company, an appealing partner. Plus, we all knew the big guys were banging at the door: companies like Intel that could potentially build or invest in similar-performing modems by brute force, hiring hundreds of engineers and plowing through the same arduous terrain we'd managed to painstakingly navigate with our small team. What nobody could foretell was how long our timing advantage would last.

The message was clear. I was the only person in the room who did not want to sell LANcity. But the commonwealth ruled. I placed a call to Severino and Sachs. I told them of our decision. We'd be privileged to join the Bay Networks family. Let's do it. The rest was paperwork and back-and-forth sharing of documents as Bay Networks rounded out the due diligence effort. One major provision was that I'd need to sink a portion of the proceeds into an escrow account to allow for any sort of intellectual property defense. I had no objection: I knew that our work was our work, unique to LANcity. Still, our originality would later be tested as a lawsuit from Applitek's founder Ash Dahod challenged the propriety of our Unilink-II protocol. We prevailed in court after a long legal battle preceded by three days of depositions from a resolute Chris Grobicki, who knew everything there was to know about the Unilink-I and Unilink-II protocols and their distinctions. Even so, the breach of a personal relationship left a scar.

So there it was. I arrived at a downtown Boston conference room on a crisp September afternoon in 1996 to seal the deal, signing my name repeatedly across a stack of five books, each four inches thick. Bay Networks had come up with the name Showtime Merger as the handle for the legal entity that would complete the deal.

Showtime indeed. Minutes afterward, the wire transfers happened, money flowing to my LANcity team, landing in bank accounts scattered around Boston. I'd also convinced Bay Networks to tuck away additional stock options to be used as an incentive for key employees to stick around.

I grew up in a home near Tehran with no running water. I rode a girl's bicycle to buy chicken bones. In 1981, working for General Electric, I was paid an annual salary of $13,000. Now, as if by an invisible hand, our LANcity team and I had achieved unquestionable financial success.

I left the building and drove back to our office in Andover, determined to see that LANcity, now a part of Bay Networks, didn't skip a beat. I knew the LANcity-Bay Networks deal would be a harbinger, the first of a series of transactions in the cable-broadband category that would prove the financial viability of the sector. Our deal was a signal to the wider world: Cable broadband was here, it was real, it would involve major players in the tech community, and it was unstoppable.

Prominent people in the cable industry cheered the Bay Networks deal because it signaled that our technology, much admired in a nascent category, would live on. Continental Cablevision's David Fellows told *Multichannel News*, "No matter how much you like LANcity, you have to realize that they're a small company—in the back of your mind, you're always wondering if they're big enough to survive." A top technology executive for Comcast, Steve Craddock, made a similar remark. "LANcity is small, and to do this, you need to surround yourself with money, manufacturing capacity and experience." They turned out to be correct. The assuredness of survival conveyed by the Bay Networks deal opened up the spigot, with major cable companies committing to large modem purchase orders. For my part, I knew I had to let go. LANcity was my child, my life's work, but I was comforted to know my offspring was in the hands of a capable and experienced parent.

Still, I knew my work wasn't done. The technology we had conceived, nurtured, and brought into adulthood was still barely a blip on the international telecommunications scene. To make good on my vision of a connected world where people could work from anywhere, and where information would travel across great distances at astonishing speed, we'd need to do much more.

It turned out I wasn't the only person thinking about bigger opportunities for cable modems. Not long before I'd sold LANcity, Continental Cablevision's Bill Schleyer had taken note of the explosion in cable modem technology that was happening before his eyes. In a meeting with the CableLabs board of directors during the cable industry's Western Cable Show in December 1995, Schleyer asked why cable companies had to pick and choose among multiple modem makers with multiple, conflicting technologies. What if, he wondered, we could standardize around a common set of technologies? With a standard technology platform, the day might arrive when consumers could choose their own cable modems. If they moved across town or to another state, the modem they carted along with them could plug into a new cable system and work just fine. What's more, standards could bring profound economic scale to the cable modem category, reducing costs and spurring faster development.

Schleyer had been a vital industry colleague and a big proponent of LANcity. He embraced our technology and our vision during the early trials around Cambridge. But he also saw the dark side of the cable industry's adoption of proprietary technologies. The bifurcation of the industry along vendor lines had long frustrated Schleyer and his peers. Swapping or acquiring cable properties was complicated because of the mismatch of platforms and protocols. So was the quest to achieve economic scale. So long as the industry was cleaved into two or more warring technology platforms, the benevolent lift of volume purchasing would be stifled.

"This is nuts, guys," Schleyer remembered saying to his CableLab peers around the conference table, including the TCI CEO John Malone. "Let's take control of this. Let's tell the vendors exactly what we want." Schleyer could tell he was making his point. Malone and the other colleagues leaned in, listening intently. They had been worried about the same thing Schleyer worried about: a promising market where, absent some sort of standardization, everyone would go spinning off in a different direction, with cable modems that worked in one market obviated in the next and with vendors able to squeeze buyers for premium prices on the basis of exclusive technology. Minutes later, after the board meeting had adjourned, Malone appeared on the stage at a standing-room-only panel discussion in an expansive convention center ballroom, floating the appeal of cable modem standardization to a room full of peers, vendors, and industry journalists. I was in the audience, listening closely. Malone's words stuck in my brain.

Now, alone in the office, months after the Western Show and immediately after the sale of LANcity, I thought back to what Schleyer had said. Standards for cable modems. Now there was an idea with promise. My own second act was about to begin.

CHAPTER 12

FATHER FIGURE

DOCSIS was the love child of a wide-ranging collective of individuals determined to put their own ding in the universe.

Codified in a document that ran 207 pages, the DOCSIS blueprint was released in March 1997 by CableLabs. DOCSIS was the love child of a wide-ranging collective of individuals determined to put their own ding in the universe. It was a team of architects who had put it down on paper, coupled with a close-knit group of cable industry engineers, CTOs, and bosses who had nurtured it to life. Technology creators and vendors—LANcity included—had offered up their innovations and discoveries to the effort.

Among these early DOCSIS architects was an ex–Air Force pilot, Robert Cruickshank. I met Cruickshank, a gregarious optimist with a ready smile, in the spring of 1995. He'd traveled to our office in Andover to attend one of our LANcity University classes taught by our training team leader Gene O'Neil, who had been promoted from our operations team, along with our right-hand man Les Borden and our sultan of Stow, Nick Signore. It was part of a fact-finding mission Cruickshank was undertaking to understand cable modem technology: what it was, how it worked, and whether it had sufficient legs to play in the real world, at scale.

We hit it off from the start, sensing within minutes of sitting down at an outdoor picnic table that we shared a devotion to technology as a means of making the world a better place. Cruickshank recalled his initial impressions of me and my endeavors in *Souls of DOCSIS*, a short-form book he wrote to describe the origins of DOCSIS: "He was kind and put me at ease," Cruickshank wrote about me. "I immediately got the sense that he wanted to help me."

That I did. Like me, Cruickshank was a lifelong tinkerer—a kid who grew up in Scarsdale, New York, working on cars and hanging out in the high school wood shop. Armed with a master of science in mechanical engineering, he'd taken a job at AT&T Bell Labs in Denver, later earning a Ph.D. from the University of Colorado at Boulder and flying planes for the Wyoming National Guard on weekends. Cruickshank began working in 1994 for a new organization with the shorthand name of CableLabs, settling into a new office looking out over the Flatirons foothills west of Denver.

CableLabs, formally Cable Television Laboratories Inc. at its founding, was the brainchild of a cable television systems owner named Richard Leghorn, a former WWII Army Air Force reconnaissance squadron commander and MIT graduate with an abiding appetite for invention. Leghorn had advocated since the early 1980s for a shared research and development agency that would bring symmetry to a host of scattered technology efforts occurring in the industry.

The idea behind CableLabs was undeniably a hive mind: If something worked well for one company, chances are it could be replicated for another—a testament, once again, to the commonalities at work across a Balkanized cable television map. So long as CableLabs steered clear of delicate pricing issues or other potentially anticompetitive activities, it had free rein to explore technologies on behalf of its members and allow for open exchange of ideas.

In the organization's early days, CableLabs pursued a mostly video-centric agenda, focusing its early efforts on high-definition television, fiber-optics technologies, and signal security issues. But around the time Cruickshank came aboard, a new idea was in the mix: the what, why, and how of delivering Internet traffic at high speeds.

For some industry leaders, it was an alien concept. During an early 1995 meeting with executives from Intel, the then-CEO of Time Warner Cable, Glenn Britt, had a question. "We literally said: 'What's the Internet?'" Britt recounted in a company history book. "We had no idea. Had never heard of it."

That would soon change. Although a main focus for Britt and Time Warner Cable was the Full Service Network experiment in a suburb of Orlando, a senior lieutenant named Carl Rosetti and other executives from Time Warner Cable began to study the possibilities of data delivery. The company's CTO Jim Chiddix and a forward-thinking colleague, Mario Vecchi, were about to emerge as two of the power players in the cable-broadband mix as Time Warner planted early seeds for the high-speed Internet service Road Runner.

As enthusiasm about data transmission rose, the hive mind got moving. Four of the largest North American cable companies, Time Warner Cable among them, banded together to explore ways to harness emerging digital technologies. Working from a nondescript office suite in a Denver suburb, they incorporated a company called Multimedia Cable Network System Partners, or MCNS. MCNS's partners—originally Comcast, Cox Communications, TCI, and Time Warner Cable (joined later by Continental Cablevision and Rogers Communications)—initially focused on efforts to develop set-top box technology, specifically a do-it-all integrated circuit and related security features. But they soon pivoted to exploring ideas for high-speed cable modem technology. From there, it didn't take long for the group's backers to combine their efforts with those of CableLabs.

The CableLabs modem evaluation effort had started out with a sea of paper. More than 40 companies, including tech giants Nortel, Siemens, and Alcatel, had replied to a 1995 request-for-proposal document concerning cable modems and high-speed data transmission. They submitted stacks of thick binders stuffed with detailed technical information—schematics, illustrations, technical descriptions. Cruickshank and his team were disappointed to realize that many of the entries depended on a technology known as Next Generation Digital Loop Carrier, a 1980s-era timesharing approach that parceled out bandwidth to every terminal attached to a network, whether the user needed those digital lanes at the time or not. Cruickshank later described the concept as "an absolutely brain-dead approach for cable TV systems."

Cruickshank and his CableLabs colleagues reasoned there had to be a better way to make use of the cable infrastructure to pump out high-speed data streams over metropolitan geographies. Looking for other approaches, they were drawn to some alternatives that had been suggested by a handful of the RFP respondents, including LANcity; the Milpitas, California, company Com21; and others. Drawing from some of these ideas, a second RFP was written in a flurry of concentrated effort by Masuma Ahmed, a CableLabs project manager and chief architect who had earned a PhD in physics from Yale University. That document, bearing the nondescript title "High Speed Cable Data Service RFP," provoked a deeper burrowing into the early machinery of the cable broadband revolution.

The CableLabs evaluation effort started humbly. Cruickshank and a small team of engineers—two were University of Colorado engineering students interning for the summer—had poked around to identify tools they could use to track the speed of data streams as they sped across a cable network. A key contributor here was a high-performance data network analyzer, a machine we knew well. Our chief technology officer, Gerry White, had demonstrated our analyzer to Cruickshank during his training excursion to Andover.

We had jumped at the chance to be included in these early evaluations, inviting CableLabs to borrow our equipment, examine our test scripts, look over our internal reports and findings—anything that might be valuable to their efforts. I was confident that based on the combination of our Unilink-II protocol and the powerful cable modem we had designed around it, we had a good chance to shine.

That's exactly what happened. Test after test, the CableLabs team ranked LANcity's modem superior to others on a range of performance metrics. One query was considered the essential question of the day: How much raw data moved over the system? The verdict: "Of the three modem systems," wrote Cruickshank, "the LANcity modem moved the most traffic successfully—by far." Another test looked at the ability of a modem system to properly reassemble data packets that had been tucked inside dedicated data frames. Here again, we stood out: "The back-to-back capability of LANcity very closely resembled the performance of a wired Ethernet connection across the full range of frame sizes," Cruickshank and team determined.

Cruickshank and his engineers also looked at latency—an indicator of how fast a network produces a response to an instruction, such as a keystroke. On this measure, too, we stood out, with the CableLabs team concluding, "LANcity performed the best, with latency never exceeding one and a half milliseconds."

There was one more interesting test. CableLabs put together a choreographed surfing routine involving a handful of early web pages—roaming from CableLabs's own home page to pages maintained by @Home, TCI, and the Canada cable company Rogers Communications. They topped the list with a page displaying a photograph of nearby Longs Peak in Colorado. Our modem system swept through the progression faster than our competitors' devices, clocking in at 108 seconds. "As expected," Cruickshank wrote, "the LANcity network performed the best, having times very close to the hard wire Ethernet network."

It was one of the first serious, third-party evaluations of a new technology that was poised to change the world. And we'd come out on top. "LANcity had a very powerful system," recalled Brian Reilly, one of the CableLabs engineers involved in the early testing. In a panel discussion marking the 20th anniversary of DOCSIS, Reilly observed, "If you clobbered it with data, it would degrade gracefully, and recover gracefully."

True enough: We'd been clobbered in the past, and we'd figured out how to come out on the other side. Now, as the cable industry began to get comfortable with the idea of cable modems connected to their infrastructure, the real action was about to begin. Not in a laboratory, but in the real world. In places like Orange County, California.

* * *

Excitement was building as cable companies began to get comfortable with the idea of an entirely new business built around cable modem technology. It wasn't just that the industry's engineers were enthused about the technology. Now, people in charge of selling cable services were starting to see the possibilities, too. Joe Rooney was one of them.

Rooney was a rising cable industry marketing professional who lived and worked in Orange County. He had learned the cable business the cold-call way: as a door-to-door sales representative in Iowa City. Later, Rooney worked as the general manager of a trio of systems in Pennsylvania that were owned by Times Mirror Co. When Atlanta-based Cox Communications acquired Times Mirror's cable operations in 1994, federal rules preventing the ownership of cable TV operations and over-the-air TV stations in the same market required that Cox sell off the Pennsylvania properties. (Cox owned a Pennsylvania TV station, the NBC affiliate KDKA-TV, and chose not to sell off the lucrative asset.) But the Cox bosses wanted to keep Rooney. They offered him a job as the marketing director for the company's large Orange County cable operation, a flagship property where Cox engineers had begun to change out line amplifiers to support two-way signaling.

Here again, techno-serendipity was at work. Like many of its peers, Cox had become excited about possibilities surrounding digital television and video-on-demand offerings and was retooling its systems to support them. But troubles in getting a new digital TV guide to work with pricey new digital set-top boxes delayed the digital TV quest. That left room for Cox to think about another new, untested service that also demanded two-way networking: delivering high-speed data traffic over cable. In the winter of 1996, as the DOCSIS development effort gained momentum, Rooney began devising a marketing plan for the introduction of Cox@Home, one of the offshoots of the high-profile @Home organization that was an entryway into the Internet delivery business.

In the old days of cable TV marketing, Rooney often relied on the mainstay medium of direct mail, with flyers and postcards blanketing nearly every home in a market to introduce new channels and television options. "Spray and pray," he called it. Here, though, Rooney had an inkling that selling an unfamiliar, fast Internet service wouldn't happen unless people experienced it first. The key would be to get them in front of a PC connected to Cox's high-speed network.

An oversized tent was just the ticket. Rooney bought one from a local supplier and festooned it with the Cox logo. Not knowing entirely what to expect, Rooney and a team of associates began to set up shop on Saturday afternoons, alternating from the parking lot of a nearby CompUSA store to nearby high school gymnasiums, where they invited people to take a look at the new $39-per-month Cox@Home Internet service. Underneath the billowing white tent, Cox's people rigged up a side-by-side demonstration: On one personal computer, a 28.8 kilobit-per-second, dial-up modem was connected to a phone line. On a second PC, the Cox@Home service roared to life at 1.5 megabits per second—more than 50 times faster than the phone modem.

Most visitors had never seen a cable modem in action. Users who were accustomed to laborious wait times in the dial-up world were wowed as web pages instantly materialized—especially members of a cohort Rooney and his associates came to privately call "BBWB" for "big boys with beards." These were prototypical early adopters—video gamers, tech enthusiasts, and, not insignificantly, people with an appetite for downloading digital music files. The mp3 digital music format was becoming popular among music fans, and the BBWBs saw high-speed data as a key enabler.

Not only was Cox@Home fast, but it was also affordable. Secondary phone lines some customers had ordered to avoid clogging up the main household line while roaming the Internet were much slower, but cost about the same amount, or more, than Cox was charging.

As the weekends rolled on, the crowds grew larger. At a Saturday afternoon demonstration at Mission Viejo High School, hundreds of visitors lined up for a chance to experience a revved-up Internet. Many signed up on the spot. Rooney preserved a recorded voicemail message from the CEO of Cox Communications, Jim Robbins, who congratulated him for signing up 50 new customers over a single weekend. "We were the pride of Cox Communications," Rooney recalled.

Although demand was strong, the early logistics were trying. A few days after a customer scheduled an installation, Cox's two-person crew—one technician who knew how to connect cable lines to the home, another a personal computer expert—would show up to begin the process of hooking up a home PC to Cox's LANcity cable modem. In the early stages, Cox could typically manage only two installations per day. Customers' computers often lacked basic hardware that technicians would have to install, in addition to the usual tasks of careful line provisioning and testing, which meant that it took almost four hours to complete a typical installation.

Regardless of the pace of the installation, Rooney was floored. Demand was much higher than even the most optimistic deployment scenarios had contemplated. Part of it was purely organic, powered by a wave of word-of-mouth endorsement from euphoric early adopters. Not many people had cable modems, but those who did were eager to sing the praises of a game-changing technology.

Rooney was quick to take advantage, seizing on a marketing technique he called the "cul-de-sac effect." Mission Viejo and dozens of nearby communities were upscale suburbs whose curvy streets were rimmed with finely tended lawns that surrounded a dead-end loop. Rooney realized that if he could get one or two homeowners on a given block to sign up, they were almost certain to boast about the joys of their new cable modem to neighbors. He hired a banking industry database marketing expert to put together a geo-targeted direct-mail campaign, mailing colorful postcards to homes edging up against these same streets. "Once you got the first two," Rooney observed, "it would double to four pretty easily."

Rooney and a few peers around the industry were starting to realize that they had lightning in a bottle. A 1996 *Wired* magazine article captured the fervor: "Anybody stuck with a 14.4-Kbps modem knows how soporifically slow Net cruising can be, especially when there are graphics, video, or sound files to download," wrote journalist Lucien Rhodes. "A 28.8-Kbps modem performs better, but not much. An ISDN line, assuming you can get one where you live, can boost access speeds to 56 Kbps. But that's chump change compared with what coaxial cable can deliver."

* * *

The appeal of fast Internet connectivity came as a welcome surprise to higher-ups. It had been the business of connecting telephones, not computers, that the cable industry had been counting on to support a rebirth. Cable companies, including Cox, Comcast, and dozens of peers, had been working to get in on the residential telephone game since the passage of the breakthrough 1996 Telecommunications Act, a far-reaching policy measure that sparked a flood of dealmaking and investment in the communications arena.

Shane Portfolio was one of the people who saw cable Internet service overtake phone line fervor from the inside out. The former US Army platoon sergeant had launched his cable industry career in 1997 at one of the entry-level rungs on the ladder—serving as a Denver-based customer service agent for TCI. Two years in, Portfolio was promoted to a newly created position supporting customer service for a novel high-speed data service that had just been introduced by TCI's successor company, AT&T Broadband. "They wanted a small team to start answering phone calls for a trial to see if this technology was actually going to work," recalled Portfolio.

It did. AT&T Broadband, like every other cable company that had begun to deploy Internet service using cable modems, was seeing big spikes in demand.

Order after order came in by telephone, with word-of-mouth acclaim spreading fast. "They quickly saw that the trial was not just successful," Portfolio said. "It was wildly successful." As in, successful enough to change the industry's own lexicon. It was 1998 when the word "broadband" officially entered the vocabulary of the cable sector, with the Denver-based cable company MediaOne adopting the corporate tagline: "This is broadband. This is the way."

The combination of broadband, video, and phone services enabled what industry leaders began calling a "triple play" of services, all provided by the same company, all flowing over the same pipe, and all aggregated on a single monthly bill. It would prove to be an enviable competitive position. Phone companies were still behind in the pay television category, and satellite TV providers like DirecTV and Dish Network faced technical limits in providing high-speed, low-latency data connectivity. To stay in the game, these two camps fused their efforts in somewhat haphazard marketing alliances. Phone companies resold DirecTV and Dish services in order to offer bundled combinations of Internet service (via phone lines) and multichannel television (via satellite dishes). But these were force-fed, inorganic offerings, nothing like cable's do-it-all triple plays flowing over a common pipe. "Providing broadband in a 'double play' bundle with cable TV, or a 'triple play' with home phone service, helped cable establish a dominant position in the broadband market," commented industry researcher Bruce Leichtman.

When MediaOne adopted its new corporate tagline, the momentum was only beginning. Over time, as broadband services proliferated and high-speed cable connectivity became the dominant US on-ramp to the information highway, almost every conceivable form of interactive communications would migrate to the broadband ecosystem: music, television, professional collaboration, health care, education, remote work, and more. The "triple play" wouldn't last forever. Instead, it would give way to something bigger. Nobody quite recognized it as the world moved toward the third millennium (and an overblown "Y2K" computer scare), but we were building the foundation for the "infinite play," where just about anything was possible.

* * *

I watched like a proud parent as our media access control protocol went from its adolescence (Unilink-II) to a fully formed adult (DOCSIS). In March 1997, the original DOCSIS specification was adopted as an official standard by the International Telecommunications Union, with the cable industry engineering organization SCTE submitting the application and stewarding the process. Besides CableLabs itself, a key player was the consulting firm Arthur D. Little, which had been chosen as a project manager for the effort in January 1996. Stuart Lipoff, an Arthur D. Little executive and technology advisor, worked closely with Michelle Kuska, the TCI director of Network Technology and, earlier, the project leader for the MCNS project, to spearhead the effort. They were working toward the publication of a detailed set of documents that, as Kuska would later describe in an IEEE-published book, "are cable-operator supported, widely available to vendors, based on working technology, and contain low barriers to entry."

These objectives were both lofty and greatly coveted. The phrase "DOCSIS-certified" would become a hotly pursued credential. If there was any lingering idea that there might still be a bric-a-brac collective of noncompatible modems, it vanished when a coterie of influential, early-to-the-game modem companies, including Bay Networks, Cisco, and Samsung, signaled their support for the CableLabs interoperability ambition in the summer of 1997.

The DOCSIS fervor would spark a flurry of dealmaking—proof that investors and executives saw big opportunities ahead. In July 1998, Northern Telecom Ltd.—Nortel, for short—agreed to pay $9.1 billion to snap up Bay Networks, which had recently acquired LANcity. A year later, Nortel spun the business into Arris Interactive, a joint venture with the cable technology maker Antec. Then, in October 2000, Nortel agreed to exchange its 81 percent stake in Arris Interactive for $325 million in cash and a 46.5 percent interest in a new publicly traded company, Arris Inc. It was a dizzying progression. I watched the technology we'd invented change hands multiple times as the DNA of LANcity continued to infiltrate a new generation of innovators, ultimately landing in the good hands of CommScope.

Although the DOCSIS specification was enshrined in the spring of 1997, it took nearly two years for the first cable modems to be blessed with a DOCSIS seal of approval. More would follow in successive testing sessions. As cable modem demand took off, and as more vendors joined the DOCSIS club, the footprint of our protocol swelled, and the notoriety of the standard rose dramatically, so much so that one of the vanguards of the streaming media industry, the RealNetworks founder Rob Glaser, would later call DOCSIS a "gift to the world" in a speech to the National Association of Broadcasters.

With the DOCSIS profile rising fast, a few of my colleagues from the Boston tech corridor called to rib me: I must be raking in a fortune from the red-hot cable modem category they'd been reading about. They had assumed I attached a licensing fee to each cable modem that was deployed. I didn't blame them for the assumption, but I shrugged off the calculation. Money wasn't the point. I wanted to see the vision I had translated into a working product get out into the world as quickly as possible. Contributing our MAC protocol was a way to jump-start the revolution.

In the same spirit, I devoted my own efforts to the cable industry's interoperability and certification agenda through work at CableLabs. Early in 1997, at the invitation of Robert Cruickshank, I joined the CableLabs team as a full-time executive consultant, spending most of my time in Colorado to help further the DOCSIS effort.

My weekly ritual was this: On Sunday afternoons, I'd drive to Boston's Logan International Airport to catch one of three back-to-back flights to Denver. I'd rent a car that evening at the Denver International Airport and begin the 90-minute commute alongside US hwy 36 to a residential community abutting the edge of Boulder, the bucolic Colorado college town. I'd remain there the rest of the week, heading home on Friday afternoon to spend time with my family, then repeat the journey anew that Sunday. One result of my schedule: the accumulation of 2.5 million frequent-flyer miles. I often became mesmerized during these flights by admiring clouds that formed and then dissipated—giant billowing mushrooms that seemed a parable for the ebb and flow of change.

For close to three years, this was life. With Cruickshank, a team of top cable industry technology executives, and hundreds of visiting engineers and industry advisors, we worked to inaugurate a process that grew to encompass multiple certification waves for dozens of DOCSIS-certified modems and companion headend systems. The certification board and its chair, David Fellows, was a "who's who" of the trade, a group of individuals dedicated to the idea that what we were doing was important work. The collective worked with harmonious camaraderie, with conference calls every Monday morning providing rhythm and ritual to our efforts.

The mood was different on the vendor side. From the start, even as we worked informally with some early technology developers, the interoperability process was cloaked in tension. I understood. These were prominent technology companies that had worked hard to produce complex machinery. And we were now putting them under the spotlight. Worry was the rule, and competitive juices ran at full blast. Many of our contenders would make sure to pack up their modems at the end of a long day of testing, guarding the devices in hotel rooms so rivals wouldn't get wind of what they looked like—or whether they worked.

These vendors understood that interoperability and modem certification would become vital to a cable industry that presented a huge new marketplace for technology. If modems all spoke the same language, then the makers of components that were tucked inside them could ratchet up their development efforts against higher sales volume. Two prominent makers of integrated circuits, Broadcom Corp. and Toshiba, keenly wanted to emerge as standard-bearers of the category. Broadcom's hard-charging CEO Henry Nicholas brought a whirlwind of energy to the cause, steadfastly courting cable industry technologists, knowing the chip side of the DOCSIS equation would be vital. I knew the same thing. I repeatedly reminded cable industry compatriots how important it was not to allow the industry to be locked into a single manufacturer that would have an iron grip on the market.

In order to be certified, a modem maker had to undergo a rigorous set of evaluation and testing hurdles. Some could be accomplished offsite, with CableLabs later affirming the results. But the big trials had to be conducted live and in person, with the DOCSIS technical team putting the devices to the test over a 100-unit test network we'd set up at the facility. We'd start on Monday morning, gathering for breakfast and coffee with a team from the vendor company in the first-floor conference room of the CableLabs building. There, we'd spell out what was to come: a week of performance exercises, measurements, and interoperability tests. The idea was to make sure that one company's modem and its sister transmission technologies complied with the entirety of the DOCSIS specifications. That way, the CableLabs member companies could be assured they weren't locking themselves into proprietary technology.

For the first six waves of testing, we were focused on basic questions: Is the protocol working? Can you accommodate critical standards relating to both the RF side and the data side of a standard modem? Can data pass from a point of origination, through your modem, and onto a connected personal computer? Can your technology work with gear from other vendors?

It wasn't until March 1999, during the seventh wave, that the first official DOCSIS seals of approval were handed down. A few months before that, the industry journalist Roger Brown sought to capture the moment in a feature article for *CED Magazine*, where Brown was the editor. "Rouzbeh Yassini stands between a rock and a hard place," wrote Brown. He reported that I was "placed in the unenviable position as a fulcrum between the nation's largest cable TV operating companies and the myriad . . . manufacturers in the race to develop interoperable cable data modems."

Brown was a trusted figure in my orbit: a smart, serious-minded, good-humored journalist who once had me doubled over laughing with a story detailing a moment during a cable industry convention when he had accidentally locked himself outside of his hotel room in the early morning hours, wearing only a towel. Brown had doggedly followed the arc of cable modem technology since its early days, and at countless in-person meetings, interviews, and chat sessions at trade shows, I had worked to help him understand its arcane world. Now, Brown had penned a lengthy article tracking my career and contributions. The headline Brown chose for the cover would stick with me for the rest of my professional life: "Rouzbeh Yassini: the Father of the Cable Modem."

I had mixed feelings about the headline. Brown had notified me in advance that it was coming. I told him it was sure to cause some consternation and resentment inside the cable industry and the allied tech community. "I don't care," Brown told me over the phone. "I'm going with it."

The headline was flattering, to be sure. But there was no single individual who gave birth to the DOCSIS revolution. As the saying goes, it took a village. At LANcity, we produced essential breakthroughs. But other companies, names like General Instrument, Intel, Com21, Terayon, Broadcom, Hewlett-Packard, Toshiba, Zenith, and others, contributed critical innovations.

Similarly, a multiperson certification committee propelled the effort, socializing the concept of interoperability within the broader cable industry. Led by its chairman David Fellows, the committee included key figures in the cable modem's march to prominence: The engineer Doug Semon from Viacom Cable and, later, Time Warner Cable; TCI's resident mathematics genius Susan Marshall; the much-admired George Hart of Rogers Cable; and other pioneering leaders like Comcast's Steve Craddock. In a 2019 presentation that I helped organize for the Cable Television Pioneers, an industry group, Fellows and I identified close to two dozen people who were among the hundreds instrumental in the success of DOCSIS. "Everybody came together on DOCSIS," commented Chris Bowick, the former Jones Intercable chief technologist who later joined Cox Communications as the company's CTO. "The industry came together, probably better than it ever had before."

Still, as Brown also pointed out, there was inevitable tension tied to the DOCSIS interoperability effort. An especially trying time came just before the 1998 winter holidays, when a few modem makers had hoped to unleash their new DOCSIS-certified boxes to an American marketplace hungry for faster Internet. Careers and promotions were on the line. But a holiday debut wasn't to be: Nobody passed the test that year. As I'd told Roger Brown, the effort to make sure our industry's cable modems were rock-solid and completely interoperable lorded above any other consideration.

It was the proper outlook, given the gravity of our undertaking. I knew from my GE days how fatal hardware recalls could be. We were dead set on making sure any modem tagged with the DOCSIS label would stand the test of time. And they have: Since the first certification of the first DOCSIS modem, there has not been a single hardware recall of a DOCSIS modem—an important accomplishment given the risks associated with having to recall and replace hardware. A failure could have compromised the entire category. Our insistence on getting it right enabled an important technology to flourish.

Ultimately, we achieved our goal. Thomson and Toshiba were the first two modem makers to earn DOCSIS-certified labels in March 1999. Others followed rapidly as we marked the start of the DOCSIS technology era. Over time, DOCSIS would become the most prolifically deployed technology undergirding the US residential broadband Internet marketplace, a position it continues to hold today. Internationally, DOCSIS-powered cable systems represent the dominant wireline broadband technology in North and South America and are prolific across much of the world, per the Organisation for Economic Co-operation and Development.

In Europe, a major instigating force was Sudhir Ispahani, the chief technology officer for the international cable company Liberty Global. Ispahani took the DOCSIS blueprint and ran with it overseas, becoming the first president of EuroCableLabs, which developed the same type of interoperability and testing regimen that had vaulted the original DOCSIS specification to prominence in the United States. Ispahani mixed a hard-charging determination with an amiable, welcoming style, serving as a willing advisor to innumerable suppliers, vendors, and advanced technologists who transformed the way Europeans connected to the Internet.

Our patience and persistence were making an impact in the marketplace, creating a model for future projects. The certification program provided a working blueprint for how interoperability of new technologies could be achieved in the global cable industry. I was proud that our LANcity team had the foresight to build "hooks" into our Unilink-II protocol that would invite ongoing innovations and iterations. We had purposefully future-proofed the cable modem revolution, leaving the door open for later enhancements.

Following my work at CableLabs, I returned after a brief transition to help out one more time in developing successor DOCSIS iterations and work on a separate home networking project. By then, in 2003, DOCSIS was well on its way, with a speedier, more robust 2.0 specification released. The specification has continued to advance ever since. By the summer of 2024, the architecture we developed for the original DOCSIS standard in 1997 had grown into adulthood, with the latest specification, DOCSIS 4.0, spelling out a framework for supporting downstream data rates of nearly 10 gigabits per second. DOCSIS has come a long way.

This reality was swirling in the back of my mind in October 2024, as LANcity and three other companies instrumental in establishing the foundation for DOCSIS were recognized with Emmy Awards from the National Association of Television Arts & Sciences. I was there to accept an Emmy in the category "pioneering technologies enabling high performance communications over cable TV systems," along with Henry Samueli, a cofounder of Broadcom Corp., and Ed Breen, the former CEO of General Instrument Corp. The electronics company 3Com also won an Emmy in the same category.

I invited more than a dozen former LANcity colleagues and their spouses with me to the Emmy presentation that evening in Manhattan in an emotional reunion—the first time many of us had been together since the acquisition of LANcity in 1996. Cheryl Ulm, spouse to John Ulm and the woman who long ago had demanded that we not go bankrupt, pulled me aside to express thanks for upholding the pledge—and making LANcity an important part of her family's achievements and financial success.

The Emmy recognition was rewarding. So was some of the post-ceremony reaction. In a social media post, the longtime cable industry engineer Tom Nagel of Comcast described our achievements as part of a "brave new world" and wrote that "Rouzbeh, almost singlehandedly, didn't allow this industry to delay that evolution." John Chapman, the former CTO of Cisco's cable division, wrote: "Congratulations, Rouzbeh, you started it all going. We stand on your shoulders." Humbling praise from individuals I admire a great deal.

But as I reminded the industry journalist Jeff Baumgartner at the ceremony, things change. The DOCSIS standard will one day give way to a different means of distributing data streams at astonishing speed: pure-play fiber-optic networks that extend all the way to a home or building, with no need for a last-mile coaxial cable bridge. After all, in terms of information delivery, nothing is faster than light. "At some point, there will be no coax left, and that will be the end of DOCSIS," I told Baumgartner in an article published in the industry journal *Light Reading*. "I think I will be happy to tell you that I'm the guy who started it and I'm the guy who will actually see the end of it."

Helping to make broadband a reality for everyday citizens was my longstanding ambition. DOCSIS was a key part of the journey. As I exited through the front door of the CableLabs building for the last time, I was certain I had held up my end of the Zoroastrian bargain: I had done good deeds.

EPILOGUE
A New Alphabet

The broadband grid is now the heartbeat of the world economy.

Nearly 40 years after I first conceived a vision of high-speed data running over residential cable television, the contribution of broadband had become immeasurable. Globally, the Organisation for Economic Co-operation and Development (OECD) indicated there were nearly 2.4 billion broadband connections humming away worldwide by 2024 that were made up of roughly 500 million "fixed" network connections (those serving a stationary location such as a home or business) and close to 1.9 billion roaming/mobile broadband subscriptions.

Within the "fixed" segment, cable systems continue to play a signature role, particularly in the United States, the birthplace of modern broadband. Here, cable remains the runaway market leader, accounting for more than 76 million connected addresses, or two-thirds of the country's total number of high-speed Internet connections as of 2024.

From a financial perspective, broadband has had a dramatic impact on the cable business. An inflection point arrived in 2022 as the industry giant Comcast reported generating $24.4 billion in annual revenue from its broadband data services, while the legacy side of the Comcast house—sales of TV channels—took in $21.3 billion. For the first time, broadband data delivery had risen to become the no. 1 annual revenue source for the world's largest cable company.

Other cable companies are on the same path. "It's really becoming much more a connectivity business, and much less a video transport business," pronounced the industry leader John Malone, then the chairman of the media and communications investment group Liberty Media, in a November 2022 CNBC interview.

From my vantage point, the business of what used to be "cable television" is already fully engaged in connectivity, creating and sustaining the anything, anywhere, anytime network. The broadband grid is now the heartbeat of the global economy, powering essential pursuits of humanity with a range that is breathtaking. In my book entitled *Planet Broadband* (2003), I pointed to more than 70 high-profile, personalized, consumer-based applications that broadband has enabled or enhanced, touching sectors ranging from sensor-infused environmental monitoring to online education experiences to collaborative work. Broadband-powered telecommuting has become an everyday reality, with more than one-fourth of the total full-time work days in the United States involving work from home, per a 2023 Stanford University survey. (The implications for fossil fuel consumption here are positive and far-reaching.) Social relationships, digital e-commerce, and the modern-day explosion in streamed entertainment are also powered by broadband, evidence of a stunning turnabout where individuals have the tools to become prolific and successful producers. Other everyday experiences, from digesting news to participating in virtual meetings to paying taxes, draw from broadband's always-on connectivity.

Beyond these examples, the entire tech sector, with Silicon Valley at the forefront, has been redefined because of broadband. Cloud computing, AI, virtual reality, self-driving cars, connected appliances, and personalized innovations that are rapidly developing—think of connected robots that serve as companions or holograms coming to life in the living room—ride over the broadband pipe.

All of these examples take advantage of what broadband makes possible: a generational shift from hardware-intensive and expensive computing infrastructures to nimble, agile application-driven innovation and services on demand. In my view, none of these innovations hold more promise than the emergence of the AI-infused, personalized, connected healthcare category, which promises to open doors to better, healthier, longer, and more enriched lives for everyday global citizens to an extent we've never believed could be possible. It was the cable industry that paved the way for a revolution in personalized entertainment. Now, the stage is set to reprise the feat in the realm of personalized healthcare, drawing from the underlying DNA of broadband.

To be sure, there are dangers lurking around our generational broadband revolution. It's naive to think bad actors aren't also empowered by broadband; they are. Large-scale social manipulation and digital intrusions into computing systems are just two areas of concern. The convergence of advanced broadband networks with artificial intelligence also looms large. The potential positive benefits of generative AI are limitless for healthcare, consumers, government, transportation, agriculture, food industry, education, military branches, and other pillars of our society. But I'm troubled about AI technology being concentrated in the hands of a few powerful companies or self-serving oligarchs. We need to be prepared for fast-rising risks around disinformation distribution, data privacy, and convulsive workforce pressures. As broadband performance elevates, so must serious investment and innovation in thwarting malfeasants. Advanced security, trust, and guardrails are more important than ever in the broadband age. So are eager, determined innovators who can outsmart the bad guys.

The impact of broadband landed with particular force in the early spring of 2020 when the Covid-19 virus all but shut down much of the world. Here was an important test of our technology. With physical transportation sharply curtailed, we needed alternative ways to conduct the world's daily business. Schoolchildren needed to learn. Office professionals needed to work from home. Restaurants needed to take orders. People everywhere, especially elders, needed desperately to stay in touch at a time of sudden and profound isolation. At the same time, local and federal governments needed to support their citizens.

As the virus bore down on the world, broadband demand soared. Tom Rutledge, then the CEO of cable's Charter Communications, talked during a 2020 financial earnings review about the "Mother's Day" phenomenon. He was alluding to a longstanding telephone industry phenomenon where, on a designated Sunday afternoon, everybody everywhere wanted to talk to mom at the same time. In the early months of 2020 and ever since, every day is Mother's Day, and the network, thanks to a steady progression of upgrades and enhancements, is able to meet the moment.

"The Internet was a bright spot during the darkest hours of 2020," wrote Tony Werner, a talented technology executive and friend who was then president of technology and product for Comcast. Werner praised broadband for "keeping hundreds of millions of people connected to work, school, entertainment, and most importantly, each other."

Broadband's resilience attracted notice. The Broadband Commission on Sustainable Development, an organization backed by the International Telecommunications Union and the United Nations offshoot UNESCO, wrote about the role broadband played in helping the world recover from the pandemic. "Never before have broadband networks and services been so vital to our health, our economic productivity, and our safety, and to keeping our economy and societies working," the commission observed. "Digital is truly the hidden hero of this unprecedented global crisis."

My own reaction was: Of course the network held up. It held up for the same reasons I knew, tracing back to the late 1980s, that our technology could be made to perform at scale, uniting large numbers of users for long distances and at fast speeds. The reasons were familiar ones: Mathematics. Physics. Electromagnetic energy. Lessons we'd learned at Rock Island Arsenal and elsewhere and applied to LANcity's crown jewel, our LCP cable modem. The usage levels may have surged, and the bits may have flowed at historical rates, but two plus two still equaled four. Covid accelerated by many years an inevitable pivot to a connected economy. But it didn't undo math.

On a broader front, there is still work to achieve something even approaching ubiquitous connectivity for every citizen. As a country, we will need our entire citizenry to be computer/Internet literate in order to lead effectively in the new world's global economy and in innovation. Our nation should set a goal of being the world's first to achieve ubiquitous broadband deployment, whether it happens through wired or wireless access networks or, more likely, both. But we're not there yet. By 2024, close to 30 million people in the United States, nearly 10 percent of the population, were not connected to broadband. We also have work ahead on the global stage, where billions of people lack access.

Beyond physical distribution, there are issues around affordability. Governments and private industry need to continue working intelligently and creatively to make sure that broadband remains within the legitimate economic reach of every citizen—always on and always safe. A research center I established at the University of New Hampshire has taken this mission of ubiquitous connectivity seriously. The UNH Broadband Center of Excellence is focused on a few critical areas, starting with mapping out broadband's true reach, given that our present mapping and intelligence are inadequate. We're examining ways to ally with service providers in pursuit of affordable broadband, addressing the gap between broadband connectivity and low-cost availability for each US zip code. And we're advocating for a more effective policy approach. My dream is that the United States will replace a patchwork governance approach by creating and maintaining a cabinet-level Office of Broadband Services to bring certainty and efficiency to the cause.

Looking back, I can identify three light-the-fuse moments along the road to making broadband service for all an everyday reality. An important inflection point occurred in 1992 with the creation and implementation of LANcity's proven MAC protocol, which would become a cornerstone element of the breakthrough technology standard called DOCSIS. A second major event happened when leaders of the cable industry, prompted by Bill Schleyer's 1995 entreaty to the CableLabs board of directors, agreed to pursue a means of achieving interoperable cable modems. That ambition hit home in March 1999, when, during a seventh wave of interoperability testing, we were able to prove that DOCSIS data traffic could flow over multiple modems incorporating multiple chipsets from multiple vendors. Decades later, during a phone conversation, Schleyer credited broadband for a wholesale industrial resurrection. "It saved the cable industry," he told me from his home in New England. "Without this, the cable industry would not have been healthy. There would have been tremendous dislocation."

To his point, I enjoyed watching a 2023 *60 Minutes* profile of a much-admired cable entrepreneur, Rocco Commisso, an energetic Italian immigrant who founded the rural-market cable company Mediacom Communications. Once a hustling kid from the Bronx who delivered pizzas and chased gigs playing accordion, Commisso pivoted from a career in investment banking to create Mediacom in July 1995, just as the broadband transformation was gaining momentum. Commisso made a smart bet on digital cable services on the cusp of a seismic transformation. By 2023, *60 Minutes* pegged Commisso's net worth at north of $8 billion, with correspondent Sharyn Alfonsi crediting the entrepreneur's "buried treasure: 600,000 miles of cable used to carry computer data through places such as the corn belt and deep south."

Not long ago, I reacquainted with John Cyr, the former DEC colleague who had seen the promise of cable broadband early on. Over a video call, we talked about the contribution of the cable modem and its technology offshoots, cornerstone instruments of our connected age and the foundation for a forthcoming generation of services we can barely begin to imagine today. We marveled about how simple it was to visit in real-time, in high-resolution splendor, across a long distance, without so much as a hitch. A far cry from AT&T's jolty, black-and-white Picturephone display at the 1964 World's Fair in New York City.

Reacquainting with my former colleague more than 30 years later was gratifying, as was a comment John made before we ended our call. The remarkable realization isn't just that broadband over cable is prolific in today's world, he pointed out. It's also knowing where it originated.

"Remember," my friend told me, "it came from, like, thirteen people."

Actually, make that 20 people: 13 employees, 6 advisors, and me. A tiny New England company that ended up at the center of the global broadband transformation. That was the team that lit the fuse for the accidental network. The team that made good on our own American dream.

ACKNOWLEDGMENTS

Completing a book is not unlike developing new technology: it's a collaborative effort. It takes a convergence of minds and talents.

My thanks, then, to some of the individuals who were instrumental in getting to the finish line. First, to the band of brothers who helped to recount, affirm, and verify a great number of technical and historical details described in these pages: Kurt Baty, Les Borden, Bill Corley, Chris Grobicki, Paul Nikolich, Nick Signore, and John Ulm. Also, much gratitude to those who gave their time to review drafts, correct numbers, chase down details, and provide encouragement: the business journalist Jeff Baumgartner; the cable industry communications professional Peggy Keegan; former curator at the Computer History Museum, Alex Bochannek; and the historian and librarian Brian Kenny of Syndeo Institute at The Cable Center, whose remarkable collection of oral history interviews and publications greatly enlivened these pages. Alongside these individuals, a shout-out of gratitude to friends and colleagues who kindly looked over drafts and gently offered their critiques: Troy Benohanian, David Fine, Scott Harris, Jan Nisbet, Karen Sumbulian, and Daniel Ward.

On the publishing front, a heartful "thank you" as well to the team at West Virginia University Press: editorial director Marguerite Avery, the sharp-eyed managing editor Kristen Bettcher, and WVU Press director Than Saffel, who believed in this book from the start. And to Michele Dillon of the University of New Hampshire, who worked devotedly to help identify a publishing pathway for this book.

And of course, to my colleague and friend, the journalist Stewart Schley, who convinced me that I had a story worth telling and helped me—with persistence, grace, and many late-night conference calls—to tell it. My unending gratitude goes to Stewart; without his dedication and resolve, this book would not have happened.

On that note, Stewart and I also would point to a body of literature contributed by talented writers and historians. Mark Robichaux's excellent biography of John Malone (*Cable Cowboy: John Malone and the Rise of the Modern Cable-TV Business*), Kara Swisher's penetrating look at the early days of AOL (*AOL.com: How Steve Case Beat Bill Gates, Nailed the Netheads, and Made Millions in the War for the Web*), Thomas P. Southwick's colorful history of the cable industry (*Distant Signals: How Cable TV Changed the World of Telecommunications*), and Patrick R. Parson's comprehensive *Blue Skies: A History of Cable Television* are among scores of published works that have been bookmarked, dog-eared, and turned to repeatedly in the writing of this book. Also deserving recognition are the unheralded chroniclers of the telecommunications business: the many excellent trade journalists from publications including *Broadcasting & Cable*, *Cable World*, *CableFAX*, *CED*, *Light Reading*, *Multichannel News*, and others. Their work remains an invaluable resource for understanding what happened when, and why.

Among industry colleagues, there are numerous individuals who helped to make this project possible. Several who stand out, and are mentioned in these pages, are Chris Bowick, John Chapman, David Fellows, Sudhir Ispahani, Jay Rolls, and Marty Weiss. Similarly, within the broadband technology sphere, it's impossible to conceive of this work happening without the contributions of Jerry Amante, John Cyr, Lynn Jones, and Yvette Kanouff.

Finally, I'd like to acknowledge the hero of this story: the cable modem. Since its introduction in 1988, the cable modem has served as the dynamic catalyst of a broadband revolution that has created millions of jobs, produced incalculable economic value, and most importantly, empowered and connected humanity, across the globe, every day.

—Rouzbeh
September 2025

IN GRATITUDE...

Heartfelt thanks to these individuals for making this story possible. Quoting the poet Rumi: "You are not a drop in the ocean. You are the entire ocean in a drop."

Nick Hamilton-Piercy * John Cyr * Robin Lavoie * Doug Jones * Dallas Clement * Karen Rose * John Kaufmann * Lloyd Jensen * Jerry Amante * Jane Ruggiero * Chet Birger * George Hart * Jay Rolls * Stephanie Mitchko * Larry O'Sullivan * Roger McGee * Rickey Luke * Chris Detsikas * Tom Olsen * Paul Chamberlin * Bill Bauer * Gene O'Neil * Levent Gun * Bob Burke * Jamie Valente * Arjang Zadeh * Joe Godas * Mary Lautman * Shlomo Rakib * Tony DiSessa * Paul Bosco * Ed Holleran * Charles Carroll * Pamela Anderson * Doug Semon * John Ulm * Dave Peabody * Dale Hokanson * Michael Adams * Terry Wright * Bill Elfers * John Malone * Kumar Surinder * Graham Sargood * Rachael DeFeo * John Hildebrand * Chris Grobicki * Tom Kolze * Daryl Striker * John Leddy * Zaki Rakib * Bob Cruickshank * John Chapman * Lynn Jones * David Woodrow * Bill Schleyer * David Fine * Pat Hansen * Greg Dargle * Mario Vecchi * Bob Wright * Steve Craddock * Mike Durant * Jim Albrycht * Kevin Mousseau * Cathleen Quigley * Carson Chen * Dennis Picker * Nick Signore * Mark Laubach * Kathy Cuddy * Ron Hranac * Susan Adams * Nomi Bergman * Marwan Fawaz * Amos Hostetter * Roger Brown * Ed O'Connell * Parvaneh Khaknejad * Tony Werner * Dick Hibbard * Tom Staniec * Ralph Brown * Walter Conroy * Chuck Carroll * Oleh Sniezko * Kip Compton * Paul Gray * Michelle Kuska * John Bavilacqua * Les Borden * Bob Miron * Rich Woundy * Alex Best * Joanne Torla * Bob Wald * Alok Sharma * Dave Unger * John Goddard * Paul Glist * Jay Vaughan * Mark Coblitz * Jesus Lopez * Tom Cloonan * Sharon Brucato * Marc Belland * Victor Hou * Steve Van Beaver * Jim Chiddix * Fred Torrisi * Karen Kaczmarski * Tom Moore * Kenneth Gould * Mark Sumner * Martin Weiss * Dick McGarry * Dave Bither * Jim Forester * Siamak Yassini * Jorge Salinger * John Eng * Lorenz Glatz * Jay Brothwell * Jim Spindler * William Luhers * Peter Percosan * Kathy Wolfe * Bill Zabor * Tom Nagel * Esteban Sandino * Frank Jones * Kevin Casey *

In Gratitude . . .

Izzy Santiago * Pamela Yassini-Fard * Bill Kostka * Chris Bowick * Rob Stoddard * Howard Pfeffer * Kevin McElearney * Dermot O'Carroll * Leslie Ellis *Wilt Hildenbrand * Pragash Pillai * Kurt Baty * Joselyn Guzman * Cindy Mazza * Jim Robbins * Henry Samueli* Will Biedron * Dale Myers * Chuck Anderson * Ed Zylka * Andy Borsa * Jerry Lampert * Dan Pike * Jason Schnitzer * Milo Medin * Steven Rose * Claude Francisque * Denis Bélanger * Oliver Van Garr * Dan Rice * Wilson Sawyer * Socorro Guzman * Bob Raiano * Kevin Leddy * Zenone Naitza * Dennis Blanchard * Sudir Isphani * Dick Green * Stewart Schley *Luisa Murcia * Mike Hayashi * Peter Bates * Craig Brinker * Dave Fellows * Jack Fijolek * Mike LaJoie * Bob Hunt * Andrew Sundelin * Sal Privitera * Kazuyoshi Ozawa * Dorothy Raymond * Stu Lipoff * Bill Corley * Richard Hertz * Susan Marshall * Steve Dukes * Shinwoong KAY * Chris Morneweck * Chris Lammers * Sid Gregory * David Gingold * Mike Sperry * Gerry White * Mom and Dad

NOTES

Chapter 1: American Dream

2 **only beginning to probe:** *A Decade of Innovation: The History of Cable Labs, 1988-1998* (Cable Television Laboratories Inc., June 4, 1998).

10 **mandate Welch had famously issued:** Jack Welch and John A. Byrne, *Straight from the Gut* (Business Plus, 2001).

Chapter 2: Wired Nation

12 **they could watch you:** Syndeo Institute at the Cable Center, Geraldine Laybourne oral history, August 2000, interviewed by Steve Nelson, https://syndeoinstitute.org/the-hauser-oral-history-project/k-l-listings/gerry-laybourne-program-hauser-project/.

13 **codified a frequency swath:** Federal Communications Commission, *Sixth Report and Order*, April 1952. The FCC ended a nearly four-year freeze on new TV station applications, expanded the range of frequencies devoted to television broadcasting, and opened the way for more than 2,000 new TV stations across the United States. The FCC, acting under authority granted via the 1934 Communications Act, made available for television broadcasting a total of 82 TV channels, 12 of them in the very high frequency (VHF) band consisting of channels 2–13 and 70 on a newly prescribed ultra-high frequency (UHF) band, with each station allocated 6 MHz of spectrum.

13 **more than 15 million US homes:** *The Cable TV Financial Databook* (Paul Kagan Associates, June 1983).

13 **"who used to knock on the doors":** Syndeo Institute at the Cable Center, Charles Townsend oral history, August 2006, interviewed by Steve Nelson, https://syndeoinstitute.org/the-hauser-oral-history-project/c-listings/ctam-panel-second-decade/.

13 **a symbolic tipping point:** *The Cable TV Financial Databook* (Paul Kagan Associates, June 1989).

14 **with votes split between:** Patrick Parsons, *Blue Skies: A History of Cable Television* (Temple University Press, 2008). The author Parsons, attempting to resolve the "who was first" question, leans toward Oregon's Leroy Parsons, noting, "Walson's story may have been the product of a memory more hopeful or romantic than accurate."

14 **raked in more than $30,000:** Jim Mott, author interview, July 2022.

15 **counted 79 cable channels:** National Cable Television Association, Cable's Story, https://www.ncta.com/cables-story#:~:text=Number%20of%20cable%20TV%20networks,1980%20to%2079%20in%201989.

15 **"Television was being cannibalized":** Ken Auletta, *Three Blind Mice: How the TV Networks Lost Their Way* (Knopf Doubleday Publishing Group, 1992).

15 **before honing his skills:** Mark Robichaux, *Cable Cowboy: John Malone and the Rise of the Modern Cable-TV Business* (John Wiley & Sons, 2002).

16 **singer named Laura Wehrsten:** Women in Cable NYC Chapter, program brochure, September 1982. The "Follies" version of the song "Tomorrow" also included these lyrics: "When we bid what we did we really meant it. We just meant it tomorrow. Not today."

16 **had promised the city fathers:** Stewart Schley, "Dolan Does It Again," *Multichannel News*, March 22, 1982.

17 **originally were dangled as bait:** Gary Kim, "Demand For I-Net Services Will Increase: Cities," Multichannel News, Sept. 1, 1992.

17 **imagine the picture quality:** Ron Hranac, author interview, July 2022.

17 **for a new "electronic highway":** Ralph Lee Smith, "Wired Nation," *The Nation*, May 18, 1970. Smith later expanded on his view in a subsequent book, also titled *The Wired Nation* (Harper Colophon Books, 1972).

18 **"It just took time.":** Ellen Cooper, author interview, August 2022.

18 **"most of you here today will be building":** James Dixon, Telecable Corporation, "The Real Work of Two-Way," *NCTA Technical Papers*, May 1972.

20 **"Draw something. Anything.":** Richard Robinson, author interview, July 2022.

21 **"Track your stocks. Educate your family.":** X*Press Information Service. A six-page promotional document described X*Press as a "free, valuable information service, direct to your PC from your cable company."

21 **at least some of their DNA:** Michael A. Banks, *On the Way to the Web: The Secret History of the Internet and Its Founders* (Michael A. Banks, 2008).

21 **"a global information market":** Kara Swisher, "Will AT&T Be Able to Make Its Web Dreams a Reality?" *Wall Street Journal*, September 25, 2000.

21 **"1980s retro-future fantasy":** Cecilia D'Anastasio, "The Secret History of a Fleeting, Pre-Internet Digital Media Channel," *VICE*, March 3, 2017.

21 **"an idea whose time has not arrived":** James Granelli, "Knight Ridder is Closing Its Videotext Service," *Los Angeles Times*, March 18, 1986.

22 **T1 lines that had been developed by Bell Labs:** In the 1960s, Bell Labs developed the T1 line—a multiplexed bundle of 24 voice channels—to support rising demand for traffic across phone networks. With guaranteed data rates of 1.54 megabits per second, they rose to commercial prominence beginning in the early 1980s for banks and other companies and organizations that required generous bandwidth to accommodate simultaneous voice calling and digital data transmission. But with monthly costs routinely exceeding $1,000, they were priced outside of the reach of residences. A later (1980s) phone network enhancement called integrated services digital network (or ISDN) was significantly cheaper, but with data rates for residential users topping out at 144 kilobits per second, provided only incremental speed advantages over normal phone lines. Critics said ISDN should stand for "innovation subscribers don't need."

22 **idea for a Prodigy-like service:** Kara Swisher, *AOL.com: How Steve Case Beat Bill Gates, Nailed the Netheads, and Made Millions in the War for the Web* (Random House, 2008). Swisher's exhaustive and entertaining chronicle of the early days of AOL ends on a positive note, suggesting that AOL had cracked the code on how to sustain a scaled online platform. Several years later, that ideal proved wrong, as AOL struggled following its 2000 merger with Time Warner Inc.

22 **"couldn't engage the industry":** Syndeo Institute at the Cable Center, Tom Elliott oral history, recorded December 12, 2002, interviewed by Rex Porter, https://syndeoinstitute.org/the-hauser-oral-history-project/e-listings/tom-elliott/.

Chapter 3: Vision Quest

24   **dominant networking technology:** Urs von Burg, *The Triumph of Ethernet* (Stanford University Press, 2001). Von Burg notes that by 1990, Ethernet had won out as the technology of choice for high-speed local area networking, displacing proprietary LANs along with the rival Token Ring architecture.

24   **scratched out a two-page memo:** Neil Weinberg, "Ethernet at 50: Bob Metcalfe Pulls Down the Turing Award," *Network World*, March 22, 2023. Metcalfe's sketch of a basic Ethernet architecture has been canonized many times over in technical periodicals and websites. The illustration conveys basic concepts of traffic flows and methods of taming network contention.

26   **snaked their way across towns:** *Kagan Cable TV Financial Databook* (Paul Kagan Associates Inc., 1990). Kagan estimated that by the close of 1987, 74 million US residences, equal to roughly three-fourths of all occupied homes counted by the US Census Bureau, could obtain cable television service—a metric industry participants referred to as "homes passed."

26   **"paid for American's broadband rollout":** Holman Jenkins, "A Federal Judge Rules Against Sports Fans—and TV Innovation," *Wall Street Journal*, August 30, 2024. The opinion column by Jenkins makes a case for linking the economic contribution of legacy cable TV channel packages with the modern broadband infrastructure. Jenkins wrote, "The traditional cable bundle has been unfairly maligned as making people pay for channels they don't watch. In fact, it allowed more tastes to be satisfied at a lower average cost than likely would otherwise have been the case. You might even say ESPN paid for America's broadband rollout."

27   **acquired in 2009:** Cisco Systems, "Cisco Announces Agreement to Acquire Starent Networks," press release, October 2009.

28   **enshrined in a February 1985:** United States Patent No. 4,500,989, US Patent & Trademark Office, February 1985.

37   **origins trace to 1660:** Ralph N. Fuller, *Stow Things: A New England Primer* (Stow Historical Society).

38   **had worked early in his career:** Nick Signore, author interview, February 2022.

Chapter 4: It's a Deal

42   **acquisition by an Australian company:** Bob Brown, "Applitek Agrees to Merge with Australian Firm," *Network World*, January 15, 1990.

42   **a polite, street-smart executive:** Allan Shen, "William Rice Elfers '71, 'Prince' business manager and longtime trustee, dies at 71," *Daily Princetonian*, December 3, 2020.

45   **946-acre slice of land:** An 1809 congressional act had set aside the Rock Island Arsenal land as a federal military reservation, although it took nearly 60 years for the first building, an impressive clock tower at the island's western tip, to be completed. Rock Island Arsenal's first commander, and the tower's designer, was the US Army Colonel and Chief Ordnance Officer C. P. Kingsbury, who played a major role in equipping Civil War soldiers with muskets and carbine rifles, as described in "Official Records of the Rebellion: Volume Eleven, Chapter 23, Part 1: Peninsular Campaign: Reports," accessed at http://www.historyofwar.org/sources/acw/officialrecords/vol011chap023part1/00006_01.html.

45   **users embraced new applications:** Dick McGarry, author interview, February 2023.

46   **introducing his technicians:** Dick Beard, author interview, February 2023.

49   **"how do you find a signal":** Chris Bowick, author interview, January 2023.

49   **all in on QPSK**: With QPSK, two data bits, forming pairs, or "symbols," are modulated—attached to a slice of electromagnetic frequency—at the same time. Their pairing becomes a "symbol" that then indicates which of four possible phases of a radio wave are being manipulated. The key number here is four: the "quadrature" in the acronym. By using a quartet of possible phases, QPSK is able to exploit bandwidth more efficiently than alternative modulation schemes while providing superior error resistance, even in compromised environments where interference can be prevalent. See Georgia Institute of Technology, "Quadrature Phase Shift Keying," https://propagation.ece.gatech.edu/ECE6390/project/Fall2010/Projects/group6/ExoBuzz/page1/page8/page8.html.

50   **custom software from scratch**: Paul Nikolich, author interview, May 2023.

Chapter 5: The Missing Link

53   **left mostly standing on the sidelines**: Edgar H. Schein, *DEC Is Dead, Long Live DEC* (Berrett-Koehler Publishers, 2004).

53   **DEC sold 69,000**: David H. Ahl, "Digital," *Creative Computing*, March 1984.

53   **"going to make much money"**: John Cyr, author interview, September 2022.

59   **"paradigm that nobody had addressed"**: John Ulm, author interview, October 2022.

60   **their names adorn**: US Patent and Trademark Office, US Patent 5,471,474, filed June 1993.

61   **"LANcity did so beautifully"**: Tom Moore, "DOCSIS Cable Center Panel," video interview, Syndeo Institute at The Cable Center, https://vimeo.com/449448301.

62   **smitten at an early age**: Chris Grobicki, author interview, September 2023.

Chapter 6: When Doves Fly

70   **legendary Fairchild Semiconductor engineer**: Leslie Berlin, *The Man behind the Microchip: Robert Noyce and the Invention of Silicon Valley* (Oxford University Press, 2006).

72   **never lost again to his dad**: Kurt Baty, author interview, May 2023.

77   **virtual simulation of our hardware**: Gerry White, author interview, November 2022.

Chapter 7: Death Star

80   **more than 50 million**: Patrick Parsons, *Blue Skies: A History of Cable Television* (Temple University Press, 2008). Parsons based the subscriber estimates on sources such as Warren Publishing's *Television & Cable Factbook* and the annual *Cable TV Financial Databook* published by Paul Kagan Associates.

80   **had plummeted to 62 percent**: David Zurawik, "1990 TV Season: Innovation Was In, but Viewers Tuned Out," *Baltimore Sun*, December 1990.

80   **freed much of the cable industry**: Cable Communications Policy Act of 1984, https://www.congress.gov/bill/98th-congress/senate-bill/66.

81   **their rates were "way too high"**: Cable World Staff, "Cable's Biggest Problem?" *Cable World*, May 1992. The article pointed out that "the cost of cable service was an issue that cropped up again and again in a national survey of cable subscribers."

81   **"sum and substance of the cable industry's political problems"**: Jeannine Aversa and Marianne Paskowski, "Mooney on S12: Rates Did Cable In," *Multichannel News*, February 1992.

82   **"Don't let him do any more damage"**: David Kline, "Infobahn Warrior," *Wired*, July 1994.

82 **later apologized:** Paul Farhi, "Malone Apologies to FCC Chief," *Washington Post*, June 1994.
82 **"building of the information superhighway":** John Higgins, "Deal of the Decade Hits the Wall," *Multichannel News*, February 1994.
82 **"regulatory terrorism":** K. C. Neel, "Rereg Double Whammy," *Cable World*, February 1994.
83 **Hartenstein on the cover:** Tom Kerver, "The Death Star Dawns," *Cablevision*, September 1993.
83 **"It was such a bust":** Char Beales, author interview, August 2022.
85 **everything you know about TV:** DirecTV television commercial, https://youtu.be/Hv0u0s_r15U?si=UenNQm8MRfB6Vy5i.
86 **"will never be the same":** Edmund L. Andrews, "A Cable Vision (or Nightmare): 500 Channels," *New York Times*, December 1992.
86 **sides of TV sets:** Stewart Schley, *Fast-Forward: Video on Demand and the Future of Television* (Genuine Article Press, 2000).
87 **had learned how to draw:** Rick Guerrero, author interview, August 2022.
88 **was early to the game:** Syndeo Institute at the Cable Center, Louis Williamson oral history, August 2015, interviewed by Leslie Ellis, https://syndeoinstitute.org/the-hauser-oral-history-project/w-z-listings/louis-williamson/.
89 **"the way the fake rocks":** Syndeo Institute at the Cable Center, Teresa Elder oral history, January 2023, interviewed by Stewart Schley, https://syndeoinstitute.org/the-hauser-oral-history-project/e-listings/teresa-elder/.
89 **busted its cable system:** Archer S. Taylor, "Jones Intercable's Alexandria Headend," *Communications Engineering & Design*, November 1995.
90 **"the hardest three months":** Cable World Staff, "Operators See 90 Days of Tough Sledding Ahead," *Cable World*, July 26, 1993.
90 **"in a conundrum":** Syndeo Institute at the Cable Center, Patty McCaskill oral history, April 2019, interviewed by Stewart Schley, https://syndeoinstitute.org/the-hauser-oral-history-project/m-o-listings/patty-mccaskill/.
91 **topped the one-million subscriber mark:** Amy Harmon, "Price War Erupts among On-Line Computer Services," *Los Angeles Times*, April 21, 1993.

Chapter 8: Puppy Love

93 **a young Matthew Broderick:** IMDb, "War Games," https://www.imdb.com/title/tt0086567/.
93 **became the modems of choice:** U.S. Robotics, Corporate Background and Information, 2004.
93 **"watching paint dry":** Edmund Andrews, "FCC Will Not Force Rate Increase for Digital Phone Lines," *New York Times*, May 1995.
95 **on the basis of time:** Kara Swisher, *AOL.com: How Steve Case Beat Bill Gates, Nailed the Netheads, and Made Millions in the War for the Web* (Random House, 2008).
95 **"part of the fabric of your life":** Jay Rolls, author interview, October 2024.
96 **"too darned expensive":** Char Beales, author interview, August 2022.
96 **phone companies could use their lines:** Edmund L. Andrews, "Ruling frees phone concerns to enter cable TV business," *The New York Times*, August 25, 1993.
97 **"Chalk it up to a dream":** Steve Lohr, "Dead End Deal: Why a Merger Collapsed," *The New York Times*, March 1994.
99 **"collegial nature of the industry":** George Hart, author interview, January 2023.
101 **hawking an aging Volkswagen:** Tom Southwick, *Distant Signals: How Cable TV Changed the World of Telecommunications* (Primedia Intertec, January 1999).

Chapter 9: Screen Wars

103 **15 percent of American households:** Robert Kominski, "Computer Use in the United States, 1989" (US Department of Commerce, Bureau of the Census, February 1991).
103 **More than half of US homes:** Camille Ryan, "Computer and Internet Use in the United States, 2016" (US Department of Commerce, Bureau of the Census, August 2018).
104 **"punch line of a bad joke"**: Mike Langberg, "Wink Gives Nudge to Interactive TV," *Chicago Tribune*, December 2000.
104 **entry-level digital cable box:** Motorola, DCT-1000 Cable Terminal Installation Manual, 1996.
105 **$3,000 per box:** Evan I. Schwartz, "People Are Supposed to Pay for This Thing?" *Wired*, July 1995.
106 **"If money is no object":** Leslie Ellis, "Many MSOs Are Wary of Digital," *Multichannel News*, November 27, 1995.
106 **"inching further from his reach"**: Mark Robichaux, *Cable Cowboy: John Malone and the Rise of the Modern Cable-TV Business* (John Wiley & Sons, 2002).
107 **"PCs are not going to go away":** Syndeo Institute at the Cable Center, Tom Elliott oral history, recorded December 12, 2002, interviewed by Rex Porter, https://syndeoinstitute.org/the-hauser-oral-history-project/e-listings/tom-elliott/.
109 **"mother could do that":** David Fellows, author interview, December 2022.
110 **"users will become addicted":** Mark Robichaux, "Cable Modems Loosen Up the Net," *Wall Street Journal*, December 1995.

Chapter 10: Rose Gardening

114 **at the center of his remedy:** Bill Corley, author interview, March 2023.
117 **most Americans, indeed, never knew:** Reed Hundt, *You Say You Want a Revolution: A Story of Information Age Politics* (Yale University Press, 2000).
118 **The students were transfixed:** Graham Sargood, author interview, May 2023.
120 **"looking toward the future"**: Martin Weiss, author interview, January 2023.
120 **aerospace companies had wandered off:** Martin Weiss's report on the findings associated with EC-Net was prepared in April 1994 for the cable industry research consortium CableLabs.
123 **"best thing that's happened":** Leslie Ellis, "MSOs May Commit $2B to Telephony," *Multichannel News*, August 1, 1994, p. 1.
124 **"kick the tire kind of outfit"**: Michael Schwartz, author interview, January 2023.
124 **"Yassini would arrive early"**: Leslie Ellis, *Kagan's Broadband Internet 2000* (Paul Kagan Associates Inc., October 1999).

Chapter 11: Wicked Fast

128 **"wide open for cable modems"**: Chuck Carroll, author interview, February 2023.
133 **Microsoft's $1 billion investment:** Microsoft Corporation, "Microsoft Invests $1 Billion in Comcast," press release, June 1997.
133 **subscriber total nearly doubled:** "@Home Losses on Target," *CNN Money*, January 1999.
134 **would unravel under the weight:** Lesley Cauley, *End of the Line: The Rise and Fall of AT&T* (Free Press, 2005). Cauley's up-close portrait of a changing AT&T illuminates a period of time when corporate giants, including AT&T and Time Warner

Inc., made enormous bets on where the broadband and digital media future was going—not all of them successful.
137 **"internal stuff happening":** Jerry Amante, author interview, September 2022.
139 **industry technologist Balan Nair:** Syndeo Institute at The Cable Center, "DOCSIS 20th Anniversary Panel," April 2017, https://cablecenter.podbean.com/e/s1-e12-docsis-20th-anniversary-panel/.
140 **object of Mossberg's scrutiny:** Walter Mossberg, "Cable Technology May Make Internet as Accessible as TV," *Wall Street Journal*, April 1996.
140 **shared a small-world moment:** Jim Carlson, author interview, October 2022.
141 **"modems take center stage":** Carl Weinschenk, "Cable Modems Take Center Stage at NCTA Convention in L.A.," *Cable World*, May 6, 1996.
141 **"literally reinventing this industry":** Cable World, "National Show '96 review," *Cable World*, May 6, 1996.
148 **"LANcity is small":** Leslie Ellis, "Big Bay Networks Swallows LANcity," *Multichannel News*, September 9, 1996.
149 **"This is nuts, guys":** Bill Schleyer, author interview, December 2022.

Chapter 12: Father Figure

151 **document that ran 207 pages:** Cable Television Laboratories Inc., "Data-over-Cable Service Interface Specifications," DOCSIS 1.0, November 2001. The original DOCSIS specification was published in March 1997 and later accepted as a global standard by the International Telecommunications Union.
152 **'What's the Internet?':** *Making Connections: Time Warner Cable and the Broadband Revolution* (Time Warner Cable, 2011).
152 **incorporated a company:** Michelle Kuska, *Cable Modems: Current Technologies and Applications* (Professional Education International, 1999).
154 **"Of the three modem systems":** Robert F. Cruickshank III, "The Souls of DOCSIS: Lehr und Kunst in Developing the World's First Standard Cable Modem and Internet Multimedia Delivery System," 2008.
157 **"cul-de-sac effect":** Joe Rooney, author interview, February 2023.
157 **overtake phone line fervor:** Syndeo Institute at the Cable Center, Shane Portfolio oral history, recorded November 1, 2022, interviewed by Stewart Schley, https://syndeoinstitute.org/the-hauser-oral-history-project/p-q-listings/shane-portfolio/.
158 **"helped cable establish a dominant position":** Bruce Leichtman, "The Changing Broadband Bundle," Leichtman Research Group, December 2022.
161 **"placed in the unenviable position":** Roger Brown, "Man of the Year: Rouzbeh Yassini," *Communications Engineering and Design*, January 1999.
164 **"didn't allow this industry":** Tom Nagel, LinkedIn, October 2024.
164 **"We stand on your shoulders":** Tom Chapman, LinkedIn, October 2024.
164 **"the end of DOCSIS":** Jeff Baumgartner, "The DNA of DOCSIS Gets Its Due with Tech Emmy," *Light Reading*, October 11, 2024.

Epilogue: A New Alphabet

165 **2.4 billion broadband connections:** Organisation for Economic Co-operation and Development, Broadband Statistics, https://www.oecd.org/en/topics/sub-issues/broadband-statistics.html.
165 **76 million connected addresses:** Leichtman Research Group, *Research Notes: Actionable Research on the Broadband, Media & Entertainment Industries*, 1Q 2024.

165 **Comcast reported generating:** Comcast Corporation, "Comcast Reports 4th Quarter 2023 Results," press release, January 2024.
165 **"more of a connectivity business":** CNBC, "Interview with Liberty Media Chairman John Malone," November 2022.
167 **"Internet was a bright spot":** Comcast Corporation, "Comcast Releases 2020 Network Performance Data, Highlighting COVID-19 Impact," press release, March 2021.
167 **"Never before have broadband networks":** Broadband Commission on Sustainable Development, "Covid-19 Crisis: Broadband Commission Agenda for Action," April 2020.
168 **close to 30 million people:** US Federal Communications Commission, *Measuring Fixed Broadband, 12th Report*, January 2023.
169 **"saved the cable industry":** Bill Schleyer, author interview, December 2022.
169 **"corn belt and deep south":** CBS News, "60 Minutes: Only in America," July 2023.

INDEX

ABC, 8, 13, 15, 80
Advanced Micro Devices (AMD), 78, 130
Advanced Networking Technologies, 53
Ahmed, Masuma, 153
Albrycht, Jim, 35, 39, 52, 56–58, 137
Amante, Jerry, 64, 135–37
AMD. *See* Advanced Micro Devices
American Television & Communications (ATC), 88
America Online (AOL), 22, 91, 95, 97, 111, 124, 132–33
amplifiers, 46–47, 98–99, 116, 155
    challenges, 68–69
    displacing, 88
    early failings of, 17
    and hybrid network, 89
    reengineering cable systems, 18, 113–14
    in Stow, Massachusetts, 38–40
Amsterdam 2000, 117
Andrews, Edmund, 93
Anstrom, Decker, 141
Antec, 159
AOL. *See* America Online
Apollo Computer, 59
Apple, 53
Apple II, 105
Apple PowerBook, 105
application-specific integrated circuit (ASIC), 70–72
Applitek Corp., 4, 147–48
    acquisition of, 41–45
    establishment of, 27–28
    investors of, 43
    new CEO of, 44
Armstrong Communications, 100
Arris Interactive, 159
Arts & Entertainment Network, 14
ASIC. *See* application-specific integrated circuit
ATC. *See* American Television & Communications
AT&T, 21, 87, 169
    Network Services, 118
    AT&T Broadband, 134, 157
Auletta, Ken, 15, 105

Battery Ventures, 145
Baty, Kurt, 70–76, 126–29
Bauer, Bill, 101
Bay Networks, 53, 146–48
Beales, Char, 84, 96
Beard, Dick, 46
Bellaire Cable TV, 87
Bell Atlantic, 82, 96–97
Bell Labs, 15
BellSouth, 143
Big 3 (broadcast TV networks). *See* ABC; CBS; NBC
bit error rate, 55
Bither, David, 71
Black, Roger, 132
Boggs, Dave, 53
Boon, Matthew, 38
Borden, Les, 63, 65, 92, 146
Borsa, Andrew, 48–51, 69
Boston College, cable modems in, 110–11
Bowick, Chris, 119, 162
Brandt, Jan, 91
Bravo, 12
Bresnan, Bill, 100
Bresnan Communications, 100
Brinker, Craig, 64
British Telecom, 128
Britt, Glenn, 152
broadband
    broadband-over-cable category, 118, 126, 134, 137
    connections, 165, 168
    data delivery, 88–89, 106, 165
    proliferation of services, 158
    subscriptions to, 165
Broadband Commission on Sustainable Development, 167
Broadcom, 161–62, 164
Brown, Roger, 161–62
Bush, George H. W., 81

Cable Act of 1992, 81
    fallout from, 89–91
    FCC interpretation of, 97
    and moment of reprieve, 117–19

Cable Alabama, 101, 142–43
cable business, 2, 90, 101, 132, 146, 155, 165
Cable Communications Policy Act of 1984, 80–81
"Cable Follies, The," 15–16
CableLabs, 2, 61, 123–24, 131, 138, 158. *See also* Data over Cable Services Interface Specification (DOCSIS)
  idea behind, 151–52
  modem evaluation effort, 153–54
  and DOCSIS, 151–64
CableNET, 124
Cable News Network. *See* CNN
cables
  coaxial, 13, 27–28, 35, 38, 40, 44, 46, 65, 69, 88–89, 110, 113, 116, 157, 164
  copper, 18, 46
  feeder, 40
Cable-Tec Expo, 98
cable television. *See also* television
  channels, 18, 82
  companies, 141
  competition among companies, 15–18
  conversion to digital, 86–87
  evolution of, 12–15
  federal government and, 80–81, 89–91
  infrastructure of, 17–18, 24, 57, 76, 116, 135, 139, 142
  primetime environment in, 80
  rate increases, 81
  regulation of, 81–82
  systems, 5, 16, 29, 32, 35, 40, 62, 124, 130–31, 152
  upgrades for, 18–20
Cable Television Administration & Marketing Society (CTAM), 84, 96
Cablevision magazine, 83
Cablevision Systems, 16–17, 83, 101, 102, 119
Cable World magazine, 81, 90, 141
California Cable & Telecommunications Association, 124, 131
Cambridge public library, 107
Carroll, Chuck, 128–29
Carlson, Jim, 140
cascade, 17
Casco Cable, 143
Case, Steve, 22
CATV. *See* community antenna television
CBS, 12–13, 15, 22, 80, 100
CED Magazine, 161
CERN. *See* European Organization for Nuclear Research
Chamberlain, Paul, 77
Channel Filtering Requirements, 49
channels, accommodating, 18–20. *See also* TV channels
ChannelWorks, 59, 134
Chapman, John, 164
Charles River Associates, 147
Charter Communications, 90, 95, 99, 133, 143, 167
Chicago Tribune, 104

Chiddix, Jim, 90–91
chips, testing, 76–79
Clinton administration, 22
CNN, 13, 38, 80, 90
CNN2, 14
coaxial cables, 13, 27–28, 35, 38, 40, 44, 46, 65, 69, 88–89, 110, 113, 116, 157, 164
Coblitz, Mark, 106
Coen, Dana, 16
Cogeco, 130
Com21, 124, 162
Comband project, GE, 19–20
Comcast, 15, 99, 101, 105–6, 121, 128, 131–34, 142, 148, 153, 157, 164–65, 167
Commisso, Rocco, 169
Commodore, 53
Commscope, 159
Communications, Warner, 15
Communications Satellite Corp., 83–84
community antenna television (CATV), 14
community multimedia networking, 137
Community Networking Update (report), 58
CommScope, 159
CompuServe, 91, 97
Computer Information Management Research Center (ASU), 120
Computer Protocol Corp., 42
computers, 5. *See also* screen wars
  Apple debut, 104–5
  challenge for in-home screen supremacy, 104
  and interactive television, 104
  mainframe, 28, 34
  networking, 57, 146
  personal, 45, 53, 104
  rising penetration of, 103–4
Con Edison, 50
Continental Cablevision, 90, 101, 104, 107–8, 128, 131, 143
Cooper, Ellen, 15–16, 18
copper cable, 18, 46
Corley, Bill, 10, 50–51, 61, 114, 126–129
Country Music Television, 12
Cox Communications, 20–21, 87, 95, 101, 119, 121, 128, 131, 155
Cox@Home, 154–57
Craddock, Steve, 162
Cruickshank, Robert, 151, 160
CTAM. *See* Cable Television Administration & Marketing Society
CVC, 22
Cyr, John, 53–55, 114–15, 169

Dahod, Ashraf, 27, 148
'D'Anastasio, Cecilia, 21
Daniels, Bill, 99
Data General Corp., 72
data networks, 107–11
Data over Cable Services Interface Specification (DOCSIS), 137–40, 169

adopting specification, 158–64
architects of, 151–52, 162
and CableLabs, 151–54
certification, 160–61
codification of, 151
DOCSIS 4.0, 163
recognition of, 164
and vendors, 160–61
DBS. *See* direct broadcast satellite; DirecTV
DEC. *See* Digital Equipment Corp.
DECWorld, 47
Defense Standard Ammunition Control System, 45
Delphi Data Services, 108
Design Circuits Inc., 76, 130
dial-up telephone modem, origins of, 93–95
digital cable TV, 83–87
digital cable box, 104–5
Digital Equipment Corp. (DEC), 35, 38–39, 43, 75–76, 79, 115, 119, 126, 146, 169
  design review meeting, 71
  exiting partnership, 138
  extension of agreement, 114
  finding strategic partner in, 52–58
  in history of computing, 53
  partnership materializing with, 63–65
  program manager for, 121
digital signal processing (DSP), 19
digital subscriber line (DSL), 97, 128
Diller, Barry, 105
direct broadcast satellite (DBS), 82. *See also* DirecTV
DirecTV, 82–83, 158
  subscriber count, 84
  TV commercials promoting, 85
  using digital technology, 85
Discovery Channel, 12, 90
DiSessa, Tony, 63–64
Dish TV, 83
DOCSIS. *See* Data over Cable Services Interface Specification
Doerr, John, 131–32
Dolan, Charles, 16, 83, 101, 102
Donatelli, Tom, 82
Douglas, McDonnell, 119–20
DSL. *See* digital subscriber line
DSP. *See* digital signal processing
Dunstan, Steven, 33
Durant, Mike, 146, 147

Eagle Communications, 142
EC-Net. *See* Electric Commerce Network
8-bit counter, 74
Elder, Teresa, 89
Electric Commerce Network (EC-Net), 119–23
Electronic Data Systems, 108
electronic highway, foundation of, 17–18
electronic mail, 94
electronic publishing. *See* teletext, emergence of
Electronics, Jerrold, 17

Elfers, William, 42–45
Ellinghaus, William, 21
Elliott, Tom, 22, 107
Ellis, Leslie, 124
Ellis, T. S., 96
Emory University, 30, 34, 39
Ergen, Charles, 83
ESPN, 38
Ethernet, 47, 53–54, 63, 110, 134, 154
  competing with, 24–25, 27–29
  connecting with, 39
  "Ethernet over TV," 59, 137
  office environment using, 58
  packets, 60–61
  prevalent connectivity of, 34–35
European Organization for Nuclear Research (CERN), 55

FCC. *See* Federal Communications Commission
f-connector, 47
FDDI. *See* fiber-distributed data interface
Federal Communications Commission (FCC), 13, 82, 85, 89–90, 96–97, 117
Fellows, David, 4, 107–11, 123, 148, 160, 162
fiber-coaxial lines, 119–20
fiber-distributed data interface (FDDI), 58
fiber optics, emergence of, 87–89
Fidelity Capital, 43
Field of Dreams, concept, 54, 100
file transfer, emergence of, 121
Fine, David, 5, 141, 171
Fisher, Shawn, 100
"fixed" network connections, 165
flooding, dealing with, 45–48
Florida Power & Light, 62
franchise territories, 16
FSN. *See* Full Service Network
Full Service Network (FSN), 91, 152. *See also* Time Warner Cable

GameLine, 22
Garlington, Dennis, 20
Gateway, 21
GE. *See* General Electric
General Electric (GE), 61, 70, 86, 124, 148, 162
  Comband project of, 19–20
  education provided by, 11
  Financial Management Program of, 24–26, 43
  working at, 9–11
General Instrument Corp., 15, 101, 104–5, 124, 137, 142, 147, 162
Giobbi, Michael, 100
Glaser, Rob, 159
going online, temporary transience of, 94–95. *See also* dial-up telephone modem, origins of
Gore, Al, 81, 82, 97, 122
Green, Richard, 123
Griffey, Ken, Jr., 129

Grobicki, Chris, 30, 60, 62, 71, 79, 114, 146
GTE (company), 30, 87, 108
Guerrero, Rick, 87–89

harmonic convergence, 23
Hart, George, 16, 99, 109, 162
Hartenstein, Eddy, 83
Hawe, William, 53–54, 67, 71
HBO, 13, 22, 24, 28, 38, 83, 92
HDTV, 84–85
Hearst, William Randolph, III, 131
Hewlett-Packard, 59, 124, 162
HFC. *See* hybrid fiber-coaxial
HGTV, 90
Higgins, John, 82
High Performance Computing Act, 122
Hildebrand, Wilton, 101–2
Hokenson, Dale, 64
Holswade, Larry, 31, 36
@Home, creation of, 129–34, 154
Honey I Shrunk the Kids (film), 34
Honeywell, 36
Hostetter, Amos, 90, 108
Hubbard, Stanley, 83
Hughes Aircraft Co., 83
Hundt, Reed, 82, 117
hybrid fiber-coaxial (HFC), 89, 119–20
hyperlinking, introduction of, 107–11

IBM, 53
IEEE. *See* Institute of Electrical and Electronics Engineers
impedance mismatches, 68
INDAX, 20–21
ingress, 49
Inouye, Daniel, 81
Institute of Electrical and Electronics Engineers (IEEE), 97
institutional networks, 16
integrated circuit (chip), 20
Intel, 63, 103, 130–31, 147, 152, 162
interactive television, 104–7
International Telecommunications Union, 158, 167
International Telemeter Corporation, 86
iPhone, 103
Iran, disintegration of, 6–8
Iran-Contra Affair, 33–35
Ispahani, Sudhir, 163

Jenkins, Holman, 26
Jermoluk, Tom, 133
Jones, Glenn, 101
Jones, Lynn, 121
Jones Intercable, 88, 99, 101, 119, 128, 140
Jones Internet Channel, 140

Kagan, Paul, 15
Kao, Charles, 87

Kaufmann, John, 39, 52–54, 56, 137
Kim, Gary, 17
Kleiner, Perkins, Caulfield, and Byers, 131
Knight Ridder Newspapers, 21
Krisbergh, Hal, 104
Kuska, Michelle, 159

LANcity, 68, 76, 78, 84, 101, 156, 159, 162–64
  advancement of, 53
  ambitions at, 94
  assembling components, 58–61
  back in picture, 59–60
  bad news for, 105–7
  bringing sense of purpose and determination to, 50
  blow to future of, 115
  Boston College as largest network of, 107–11
  cable modem implementation, 95, 118–21
  emergence of, 31, 44–46
  exploring possible deal with DEC, 54–58
  integrated circuit development team, 71–74
  leapfrogging to LCP, 126–29
  major theme of, 124–25
  as outsider, 99
  putting modems to test, 123
  ranking modem of, 154
  selling, 140–50
  signature of, 127
  turning point, 137–40
Langberg, Mike, 104
Lautman, Mary, 137
Laybourne, Geraldine, 12, 15, 144
LCB modem, naming, 59
LCP modem, 126–29
legacy television, 12–13
Leghorn, Richard, 152
Leichtman, Bruce, 158
Levin, Gerald, 91
levees. *See* flooding, dealing with
Liberty Latin America, 139
Liberty Media, 165
Lifetime, 14
Light Reading, 164
Lipoff, Stuart, 159
Little, Arthur D., 158–59
local area networking, 23
location, solving problem of, 45–48
Lohr, Steve, 97
loopback, testing, 77
Los Angeles Times, 21

MAC. *See* media access control
MAC protocol, implementing
  application-specific integrated circuit, 70–72
  chip testing, 76–79
  digital RF architecture, 69–70
  operational challenges, 67–69
Magness, Bob, 15

Malone, John, 15, 84, 96–97, 118, 149, 165
Marcus Communications, 133
market, seeding, 98–102
Markey, Edward, 81
Marshall, Susan, 153
Martin, Laura, 96
Matrix Partners, 145
Mazza, Cindy, 60, 64, 77–78, 146
McCaskill, Patty, 90
McDonald, Jim, 144
McGarry, Dick, 45–46
MCI, 87
MCNS. *See* Multimedia Cable Network System Partners
media access control (MAC), 32
Mediacom Communications, 169
Media General, 128
MediaOne Broadband, 88–89, 158
Medin, Milo, 132
megahertz (MHz), 13, 19, 34–35, 39, 46, 49, 61, 68, 86, 105, 108, 127, 134
Metcalfe, Robert, 24, 53–54
Metzenbaum, Howard, 81
Meyers, Dale, 48, 62
MHz. *See* megahertz
Microsoft Corporation, 105
million-node network, 124–25
MITRE Corp., 38
Monitor Channel, The, 92
Mooney, James, 81
Moore, Tom, 61
Mossberg, Walter, 140–41
Motion Picture Experts Group (MPEG), 85
Motorola, 105
Mott, Jim, 14
MPEG. *See* Motion Picture Experts Group
MTV: Music Television, 12
Multichannel News, 81–82, 106, 148
multimedia breakthrough, hope for, 20–22
Multimedia Cable Network System Partners (MCNS), 152–53
multimedia services, 2
multiprotocol communications servers, 42
MUSE system, adopting, 84–85

Nashoba Valley Cable, 37–40, 57, 102
Nation, The, 17–18
National Cable Television Association, 14–15
National Television Standards Committee, 84
NBC, 12–13, 15, 155
near video on demand, 90–91.
Needham & Co., 96
networked city, idea of, 2
network interface, term, 27
network of networks, 55
Network World, 42
Newchannels Corp., 121
New Hampshire House of Representatives, 51
news channels, emergence of, 14–15
New Yorker, The, 105
New York Times, 86, 93, 97

Next Big Thing. *See* communications sectors, convergence of
Nicholas, Henry, 161
Nickelodeon, channel, 12–13, 90
NI-10E. *See* cable modem
Nikolich, Paul, 32, 49–50, 61, 63, 69, 114
nodes, new network design in, 88–89
noise, cable television and, 17
Nortel, 53
Noyce, Robert, 70
Nynex, 30

O'Connell, Ed, 48, 60, 64–65, 77, 146
OECD. *See* Organisation for Economic Co-operation and Development
OEM. *See* original equipment manufacturer
Olsen, Ken, 5, 136
on demand, presenting, 86
O'Neill, Gene, 127
Organisation for Economic Co-operation and Development (OECD), 165
original equipment manufacturer (OEM), 58–61

Palmer, Robert, 136
Palo Alto Research Center, 24
Papantonis, Nick, 36
Parish, Robert, 139–40
Parsons, Leroy "Ed," 14
Paul Kagan Associates, 119
PC Magazine, 129
Performance Systems International, 131
Perot, Ross, 108
Picturephone, 169
Planet Broadband (Yassini-Fard), 166
Pleasure Island, 29
Portfolio, Shane, 157–58
Primestar, 83
Privitera, Sal, 64
Project Agora, 110
Proteon, 23–27, 145
punch card, era of, 8

QPSK. *See* quadrature phase shift keying
quadrature phase shift keying (QPSK), 48–51

radio frequency (RF), 10, 69–70
RAM. *See* random access memory
random access memory (RAM), 104–5
rapid, detailed, automated environment (RDA), 116
RDA. *See* rapid, detailed, automated environment
Reagan administration. *See* Iran-Contra Affair
"Real World of Two-Way, The," 18
regulations. *See also* Cable Act of 1992
    moment of reprieve, 117–19
    rate regulation, 80–81, 90, 117
Reilly, Brian, 154

1952 Report and Order, FCC, 13
retransmission consent, 90. *See also* Cable Act of 1992
RF. *See* radio frequency
Rhodes, Lucien, 157
Riano, Bob, 126
Road Runner. *See* @Home, creation of
Robbins, Jim, 101, 121
Roberts, Brian, 105, 121
Roberts, Ralph, 15
Robichaux, Mark, 107
Robinson, Richard, 20–21
Rock Island Arsenal, 28, 34, 45–48, 69
Rogers Cable/Communications, 99, 106
Rogers, Ginger, 86
Rolls, Jay, 95
Rooney, Joe, 154–57
Ropes & Gray, 4
Rutledge, Tom, 167

Sachs, Bruce, 147
Salwen, Howard, 24
Sargood, Graham, 117–19
Scheider, Roy, 144
Schleyer, Bill, 108, 149
Schwartz, Michael, 124
Scientific-Atlanta Inc., 101, 105, 137, 144–45, 147
screen wars
  establishing data networks, 107–11
  foiled dream of interactive television, 103–7
  rising computer penetration, 103–7
Scricco, Francis ("Fran"), 24
SCTE, 99, 100–101, 124, 126, 141, 158
Semon, Doug, 134
70-channel cable system, 19
Severino, Paul, 146
Shaw, Bernard, 80
Shaw Communications, 130
Sie, John, 84–85
signal trace, term, 35
Signore, Nick, 56–57, 66, 112, 146
Silicon Graphics, 133
Silicon Valley, 119, 131, 133, 146, 166
simple network management protocol (SNMP), 61
simulation, using, 75
60 Minutes, 169
Smith, Ralph Lee, 17–18
Smith, Raymond, 97
sneakernet, 58
SNMP. *See* simple network management protocol
Society of Cable Television Engineers, 98
Sorhaug, Asbjorn, 41–42
Souls of DOCSIS (Cruickshank), 151
Sperry, Mike, 64, 76, 127, 146
Spindler, James, 147
Starent Networks, 27
Storer Communications, 14

Stow, Massachusetts, 37–40, 112, 116. *See also* Nashoba Valley Cable TV
Stratus Computing, 72–73
Sun Microsystems, 59, 71
Syndeo Institute, 139

Tamagotchi, 103
TCI. *See* Tele-Communications Inc.
TDM. *See* time division multiplexing
TeleCable, 18
Tele-Communications Inc. (TCI), 15–16, 84, 86, 92, 101, 127–28, 142, 153–54, 157, 159
  backing from, 21
  and business venture, 131
  and FCC, 82
  and @Home, 132–34
  merger bid, 96–97
  potential merger with Bell Atlantic, 96–97
  successor of, 134
  and Telewest deployment, 118–19
telecommuting, 3–4
Teleport Communications Group, 120
teletext, emergence of, 20–22
television
  cable TV connections, 13–4
  collective of leaders in, 15
  competition among companies, 15–18
  evolution of, 12–15
Telewest, 117–19, 129
Tempe Precision, 120
Terayon Systems, 124, 142, 162
Texscan Corp., 101
35-channel model, 18–20
Thomson, 11, 163
Three Blind Mice (Auletta), 15
3Com, 27
tilt (phenomenon), 112
  confronting problem of, 113–17
time division multiplexing (TDM), 60–61
Times Mirror Cable, 21, 87, 119–23, 155
Time Warner Cable, 90–91, 142, 152
Token Ring, 24–27, 28
Topol, Sid, 144–45
Toshiba, 162, 163
Townsend, Charles, 13
Trintex, 22
"triple play" of services, promoting, 157–58
truck chasers, term, 13
truck rolls, 82–83
Turner, Ted, 14, 144
TV channels, 13, 15–16, 19, 26, 38, 68, 85, 89–90, 96–99, 112–13, 135, 165

Ulm, John, 54, 59, 67, 73, 114, 126–29
UNH Broadband Center of Excellence, 168
Unilink, 28–29, 34
Unilink-II, 29, 112, 158
  development of, 58–61
  protocol, 60, 66, 70, 73, 79, 138, 148, 163

United Cable Television, 13
University of Michigan, 34
U.S. Robotics, 142, 144–45, 147, 933
USA Network, 12
US Communications Act of 1934, rewriting, 96
US Patent 5,608,728, 116

Vandenberg Air Force Base, 34
VAX "super-mini" computers, 53
vertical blanking interval, 21–22
VH1, 14
Viacom, 128
Viacom Cable, 16
Vialogic, 147
video compression, 85
videotex. *See* teletext, emergence of
Videotron, 129, 130
Voom, 83

Wall Street Journal, 26, 107, 110, 146
Walson, John, 14
Walters, Bud, 47
Wangberg, Larry, 122
Warner Amex Cable Communications, 16
Warner Communications, 12
Weather Channel, The, 12
WebTV, Microsoft, 104, 106
Wehrsten, Laura, 16, 18
Weiss, Martin, 120
Welch, Jack, 10–11

Werner, Tony, 167
Wertheim Schroder & Co., 82
Western Cable Show, 86, 149
West Liberty State College, 7
White, Gerry, 60, 77, 146, 153
"wicked fast," phrase. *See* communications sectors, convergence of
Williamson, Louis, 88
Winky Dink and You, 104
Wired magazine, 157
Wired Nation: Cable TV: The Electronic Communications Highway, The (Smith), 18
wired schools, 16
Women in Cable, 15–16
WorldGate, 104
worldwide wait, term, 93
World Wide Web, 16, 93, 107
Wright-Patterson Air Force Base, 38

Xerox, 24–25
X*Press Information Services, 21–22

Yahoo!, 133
York, Barbara, 92

Zenith Electronics, 124, 142, 147, 162
Zenith personal computers, 45
Zoroastrianism, 23, 164

ABOUT THE AUTHORS

Rouzbeh Yassini-Fard is a technology entrepreneur and philanthropist whose Massachusetts-based organization, YAS Foundation, supports innovation and creativity addressing medical technology, telecommunications advocacy, educational scholarship and cultural collaboration. Yassini is known as the "father of the cable modem," tracing to the breakthrough achievements around high-speed data technologies pioneered by LANcity, the company Yassini founded in 1988.

Stewart Schley is a business journalist who specializes in telecommunications and media. His work has appeared in publications including *Variety*, *Multichannel News* and *Cable World*, where he was editor in chief. He has worked as an industry analyst for the firms One Touch Intelligence and Paul Kagan Associates, and is the author of the book *Fast Forward: Video on Demand and the Future of Television* (Genuine Article Press, 2000).